FEDERAL EXECUTIVE TEAM

Director, Climate Change Science Program: ....William J. Brennan

Director, Climate Change Science Program Office: ....Peter A. Schultz

Lead Agency Principal Representative to CCSP;Senior Advisor
for Global Change Programs, U.S. Geological Survey: ....Thomas R. Armstrong

Product Lead; Biologist, Terrestrial, Freshwater, and Marine
Ecosystems Program, U.S. Geological Survey:......................................Colleen W. Charles

Synthesis and Assessment Product Advisory Group Chair;
Associate Director, U.S. Environmental Protection Agency
National Center for Environmental Assessment.....................................Michael W. Slimak

Synthesis and Assessment Product Coordinator,
Climate Change Science Program Office: ....Fabien J.G. Laurier

EDITORIAL AND PRODUCTION TEAM

Techinical editors, U.S. Geological Survey ....Anna Glover
                                                                                          Jeanette Ishee
                                                                                          Iris Collies

Thresholds of Climate Change in Ecosystems

Synthesis and Assessment Product 4.2
Report by the U.S. Climate Change Science Program
and the Subcommittee on Global Change Research

EDITED BY:
Colleen W. Charles

January 2009,

Members of Congress:

On behalf of the National Science and Technology Council, the U.S. Climate Change Science Program (CCSP) is pleased to transmit to the President and the Congress this Synthesis and Assessment Product (SAP) *Thresholds of Climate Change in Ecosystems*. This is part of a series of 21 SAPs produced by the CCSP aimed at providing current assessments of climate change science to inform public debate, policy, and operational decisions. These reports are also intended to help the CCSP develop future program research priorities.

The CCSP's guiding vision is to provide the Nation and the global community with the science-based knowledge needed to manage the risks and capture the opportunities associated with climate and related environmental changes. The SAPs are important steps toward achieving that vision and help to translate the CCSP's extensive observational and research database into informational tools that directly address key questions being asked of the research community.

This SAP addresses and synthesizes the current state of scientific understanding regarding potential abrupt state changes or regime shifts in ecosystems in response to climate change. It was developed in accordance with the Guidelines for Producing CCSP SAPs, the Information Quality Act (Section 515 of the Treasury and General Government Appropriations Act for Fiscal Year 2001 (Public Law 106-554)), and the guidelines issued by the Department of Interior and the U.S. Geological Survey pursuant to Section 515.

We commend the report's authors for both the thorough nature of their work and their adherence to an inclusive review process.

Sincerely,

Carlos M. Gutierrez
Secretary of Commerce
Chair, Committee on Climate Change
Science and Technology Integration

Samuel W. Bodman
Secretary of Energy
Vice Chair, Committee on Climate
Change Science and Technology
Integration

John H. Marburger III
Director, Office of Science and
Technology Policy
Executive Director, Committee
on Climate Change Science and
Technology Integration

TABLE OF CONTENTS

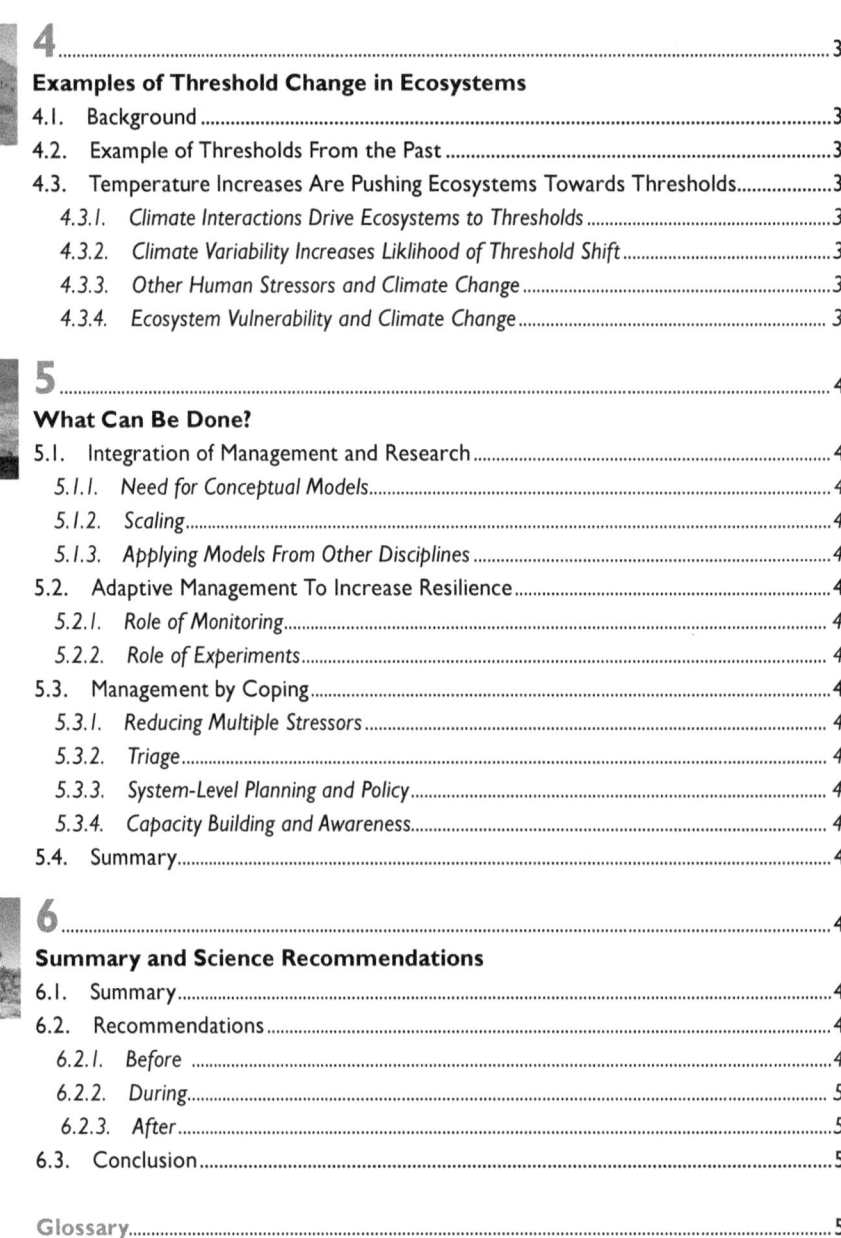

AUTHOR TEAM FOR THIS REPORT

Lead Authors: Daniel B. Fagre, USGS; Colleen W. Charles, USGS.

Authors: Craig D. Allen, USGS; Charles Birkeland, USGS-University of Hawaii; F. Stuart Chapin, III, University of Alaska; Peter M. Groffman, Institute of Ecosystem Studies; Glenn R. Guntenspergen, USGS; Alan K. Knapp, Colorado State University; A. David McGuire, USGS-University of Alaska; Patrick J. Mulholland, Oak Ridge National Laboratory; Debra P.C. Peters, USDA Agricultural Research Service; Daniel D. Roby, USGS-Oregon State University; George Sugihara, Scripps Institute of Oceanography and University of California at San Diego.

Contributing Authors: Brandon Bestelmeyer, Jornada Basin LTER; Julio L. Betancourt, USGS; Jeffrey E. Herrick, Jornada Basin LTER

Thresholds of Climate Change in Ecosystems Federal Advisory Committee

An eight-member Federal Advisory Committee oversaw the scientific review and development of this Synthesis and Assessment Product (SAP) at the request of the U.S. Geological Survey.

Linda Joyce, Forest Service

Steven McNulty, Forest Service

Ronald Neilson, Forest Service

Dennis Ojima, John H. Heinz III Center for Science, Economics and the Environment

David Peterson, Forest Service

Franklin Schwing, National Oceanic and Atmospheric Administration Fisheries Service

John P. Smol, Queen's University, Kingston, Ontario

Leigh Welling, National Park Service

ACKNOWLEDGEMENTS

CCSP Synthesis and Assessment Product 4.2 (SAP 4.2) was developed with the benefit of a scientifically rigorous first draft peer review conducted by a committee appointed under the Federal Advisory Committee Act—Thresholds of Climate Change in Ecosystems Federal Advisory Committee (FAC). The resultant review was instrumental in shaping the final version of SAP 4.2 and in improving its completeness, sharpening its focus, and communicating its conclusions and recommendations.

We would like to acknowledge Susan Haseltine, Associate Director for Biology, USGS; Robert Szaro, Chief Scientist for Biology, USGS (retired); Bruce Jones, Chief Scientist for Biology, and Jeffrey Kershner, Northern Rocky Mountain Science Center, USGS, for their support, encouragement, and accommodation provided in order to carry out this complex task.

The authors extend their sincere thanks and appreciation to Fabien Laurier, CCSP, for his patience, guidance, and encouragement all through the development of SAP 4.2.

We wish to thank the members of the Review FAC: Linda Joyce, USDA Forest Service; Steven McNulty, USDA Forest Service; Ronald Neilson, USDA Forest Service; Dennis Ojima, The H. John Heinz III Center for Science, Economics and the Environment; David Peterson, USDA Forest Service; Franklin Schwing, National Oceanic and Atmospheric Administration Fisheries Service; John P. Smol, Queen's University, Kingston, Ontario; Leigh Welling, National Park Service.

We would like to thank Anthony C. Janetos, Joint Global Change Research Institute, for providing guidance and support in developing this document and encouragement in getting this information out. We also extend appreciation to Patricia Jellison, USGS, for her support in completing SAP 4.2.

The authors also express deep gratitude to Lindsey Bengtson, Kimberly Dueling, Jeannete Ishee, and Anna Glover, USGS, for their invaluable time and dedication to the production of SAP 4.2.

We are also grateful to all the reviewers who provided valuable comments during the public comment period. The author team carefully considered all comments and the subsequent revisions resulted in improving the quality and clarity of this report.

As defined in this Synthesis and Assessment Report,

> 'an ecological threshold is the point at which there is an abrupt
> change in an ecosystem quality, property, or phenomenon, or
> where small changes in one or more external conditions produce
> large and persistent responses in an ecosystem'.

Ecological thresholds occur when external factors, positive feedbacks, or nonlinear instabilities in a system cause changes to propagate in a domino-like fashion that are potentially irreversible. This report reviews threshold changes in North American ecosystems
that are potentially induced by climatic change and addresses the significant challenges
these threshold crossings impose on resource and land managers. Sudden changes to
ecosystems and the goods and services they provide are not well understood, but they
are extremely important if natural resource managers are to succeed in developing adaptation strategies in a changing world.

The report provides an overview of what is known about ecological thresholds and
where they are likely to occur. It also identifies those areas where research is most
needed to improve knowledge and understand the uncertainties regarding them. The
report suggests a suite of potential actions that land and resource managers could use to
improve the likelihood of success for the resources they manage, even under conditions
of incomplete understanding of what drives thresholds of change and when changes will
occur.

Key examples of climate-induced threshold changes are presented. This synthesis effort
identified a suite of potential actions that, taken together or separately, can begin to improve the understanding of thresholds and increase the likelihood of success in developing management and adaptation strategies in a changing climate, before, during, and after
thresholds are crossed. In general, it is essential to increase the resilience of ecosystems
and thus to slow or prevent the crossing of thresholds; to identify early warning signals of
impending threshold changes; and to employ adaptive management strategies to deal with
new conditions, new successional trajectories, and new combinations of species.

RECOMMENDED CITATIONS

Entire Report:

CCSP, 2009. *Thresholds of Climate Change in Ecosystems.* **Fagre, D.B., C.W. Charles, C.D. Allen, Charles Birkeland, F.S. Chapin III, P.M. Groffman, G.R. Guntenspergen, A.K. Knapp, A.D. McGuire, P.J. Mulholland, D.P.C. Peters, D.D. Roby, and George Sugihara. A Report by the U.S. Climate Change Science Program and the Subcommittee on Global Change Research. U.S. Geological Survey, Reston, VA.**

Preface:

Fagre, D.B., Charles, C.W., Allen, C.D., Birkeland, Charles, Chapin, F.S., III, Groffman, P.M., Guntenspergen, G.R., Knapp, A.K., McGuire, A.D., Mulholland, P.J., Peters, D.P.C., Roby, D.D., and Sugihara, George, 2009. Preface. In: *Thresholds of Climate Change in Ecosystems.* **Fagre, D.B., C.W. Charles, C.D. Allen, Charles Birkeland, F.S. Chapin III, P.M. Groffman, G.R. Guntenspergen, A.K. Knapp, A.D. McGuire, P.J. Mulholland, D.P.C. Peters, D.D. Roby, and George Sugihara. A Report by the U.S. Climate Change Science Program and the Subcommittee on Global Change Research. U.S. Geological Survey, Reston, VA.**

Executive Summary:

Fagre, D.B., Charles, C.W., Allen, C.D., Birkeland, Charles, Chapin, F.S., III, Groffman, P.M., Guntenspergen, G.R., Knapp, A.K., McGuire, A.D., Mulholland, P.J., Peters, D.P.C., Roby, D.D., and Sugihara, George, 2009. Executive Summary. In: *Thresholds of Climate Change in Ecosystems.* **Fagre, D.B., C.W. Charles, C.D. Allen, Charles Birkeland, F.S. Chapin III, P.M. Groffman, G.R. Guntenspergen, A.K. Knapp, A.D. McGuire, P.J. Mulholland, D.P.C. Peters, D.D. Roby, and George Sugihara. A Report by the U.S. Climate Change Science Program and the Subcommittee on Global Change Research. U.S. Geological Survey, Reston, VA.**

Chapter 1:

Fagre, D.B., Charles, C.W., Allen, C.D., Birkeland, Charles, Chapin, F.S., III, Groffman, P.M., Guntenspergen, G.R., Knapp, A.K., McGuire, A.D., Mulholland, P.J., Peters, D.P.C., Roby, D.D., and Sugihara, George, 2009. Introduction and Background. In: *Thresholds of Climate Change in Ecosystems.* **Fagre, D.B., C.W. Charles, C.D. Allen, Charles Birkeland, F.S. Chapin III, P.M. Groffman, G.R. Guntenspergen, A.K. Knapp, A.D. McGuire, P.J. Mulholland, D.P.C. Peters, D.D. Roby, and George Sugihara. A Report by the U.S. Climate Change Science Program and the Subcommittee on Global Change Research. U.S. Geological Survey, Reston, VA.**

Chapter 2:

Fagre, D.B., Charles, C.W., Allen, C.D., Birkeland, Charles, Chapin, F.S., III, Groffman, P.M., Guntenspergen, G.R., Knapp, A.K., McGuire, A.D., Mulholland, P.J., Peters, D.P.C., Roby, D.D., and Sugihara, George, 2009. Ecological Thresholds. In: *Thresholds of Climate Change in Ecosystems.* **Fagre, D.B., C.W. Charles, C.D. Allen, Charles Birkeland, F.S. Chapin III, P.M. Groffman, G.R. Guntenspergen, A.K. Knapp, A.D. McGuire, P.J. Mulholland, D.P.C. Peters, D.D. Roby, and George Sugihara. A Report by the U.S. Climate Change Science Program and the Subcommittee on Global Change Research. U.S. Geological Survey, Reston, VA.**

Chapter 3:

Fagre, D.B., Charles, C.W., Allen, C.D., Birkeland, Charles, Chapin, F.S., III, Groffman, P.M., Guntenspergen, G.R., Knapp, A.K., McGuire, A.D., Mulholland, P.J., Peters, D.P.C., Roby, D.D., and Sugihara, George, 2009. Case Studies. In: *Thresholds of Climate Change in Ecosystems.* **Fagre, D.B., C.W. Charles, C.D. Allen, Charles Birkeland, F.S. Chapin III, P.M. Groffman, G.R. Guntenspergen, A.K. Knapp, A.D. McGuire, P.J. Mulholland, D.P.C. Peters, D.D. Roby, and George Sugihara. A Report by the U.S. Climate Change Science Program and the Subcommittee on Global Change Research. U.S. Geological Survey, Reston, VA.**

Chapter 4:

Fagre, D.B., Charles, C.W., Allen, C.D., Birkeland, Charles, Chapin, F.S., III, Groffman, P.M., Guntenspergen, G.R., Knapp, A.K., McGuire, A.D., Mulholland, P.J., Peters, D.P.C., Roby, D.D., and Sugihara, George, 2009. Examples of Threshold Change in Ecosystems. In: *Thresholds of Climate Change in Ecosystems*. Fagre, D.B., C.W. Charles, C.D. Allen, Charles Birkeland, F.S. Chapin III, P.M. Groffman, G.R. Guntenspergen, A.K. Knapp, A.D. McGuire, P.J. Mulholland, D.P.C. Peters, D.D. Roby, and George Sugihara. A Report by the U.S. Climate Change Science Program and the Subcommittee on Global Change Research. U.S. Geological Survey, Reston, VA.

Chapter 5:

Fagre, D.B., Charles, C.W., Allen, C.D., Birkeland, Charles, Chapin, F.S., III, Groffman, P.M., Guntenspergen, G.R., Knapp, A.K., McGuire, A.D., Mulholland, P.J., Peters, D.P.C., Roby, D.D., and Sugihara, George, 2009. What Can Be Done?. In: *Thresholds of Climate Change in Ecosystems*. Fagre, D.B., C.W. Charles, C.D. Allen, Charles Birkeland, F.S. Chapin III, P.M. Groffman, G.R. Guntenspergen, A.K. Knapp, A.D. McGuire, P.J. Mulholland, D.P.C. Peters, D.D. Roby, and George Sugihara. A Report by the U.S. Climate Change Science Program and the Subcommittee on Global Change Research. U.S. Geological Survey, Reston, VA.

Chapter 6:

Fagre, D.B., Charles, C.W., Allen, C.D., Birkeland, Charles, Chapin, F.S., III, Groffman, P.M., Guntenspergen, G.R., Knapp, A.K., McGuire, A.D., Mulholland, P.J., Peters, D.P.C., Roby, D.D., and Sugihara, George, 2009. Summary and Science Recommendations. In: *Thresholds of Climate Change in Ecosystems*. Fagre, D.B., C.W. Charles, C.D. Allen, Charles Birkeland, F.S. Chapin III, P.M. Groffman, G.R. Guntenspergen, A.K. Knapp, A.D. McGuire, P.J. Mulholland, D.P.C. Peters, D.D. Roby, and George Sugihara. A Report by the U.S. Climate Change Science Program and the Subcommittee on Global Change Research. U.S. Geological Survey, Reston, VA.

PREFACE

Report Motivation and Guidance for Using this Synthesis/Assessment Report

Authors:

Colleen W. Charles, U.S. Geological Survey

A primary objective of the U.S. Climate Change Science Program (CCSP) is to provide the best possible scientific information to support public discussion, and government and private sector decision making on key climate-related issues. To help meet this objective, the CCSP has identified 21 Synthesis and Assessment Products (SAPs) that address its highest priority research, observational, and decision-support needs. SAP 4.2 Thresholds of Climate Change in Ecosystems is one of seven SAPs developed to address goal 4 of the CCSP Strategic Plan: Understand the sensitivity and adaptability of different natural and managed ecosystems and human systems to climate change and related global changes.

In the ongoing discussions of climate change effects on ecosystems, increasing focus is being placed on the existence and likelihood of abrupt state changes or threshold responses in the structure and functioning of ecosystems. Various interrelated terms are employed in the scientific literature to characterize these types of discontinuous and rapid changes in ecosystems, including ecosystem tipping points, regime shifts, threshold responses, alternative or multiple stable states, and abrupt state changes. Such discontinuities in ecosystems are difficult to predict, and are likely to result in profound changes to natural resources that are sensitive to climate changes, as well as to human societies that depend on ecosystem goods and services. The occurrence of threshold or abrupt changes in ecosystems are suggested by current ecological theory and models, and are documented in the paleoecological record; however, they are poorly understood quantitatively as well as in terms of the underlying causal mechanisms. It is unclear under what circumstances climate change, both in its mean state and in its variance in space and time, including occurrence of extreme weather events, might cause ecosystem threshold shifts, instead of more gradual, continuous changes in ecosystems and species.

Over the past several decades, numerous scientific publications and reports have described and discussed historical and potential effects of climate change and variability on ecosystems and their constituent biota and processes. Because temperature, precipitation, and related climate variables are fundamental regulators of biological processes, it is reasonable to expect that significant human-induced changes in the climate system will have measurable effects on the distribution, condition, composition, structure, and functioning of ecosystems. Such changes in ecosystems may, in turn, alter linkages and feedbacks between ecosystems and regional climate systems. Because ecosystems produce a wide array of goods and services valued by humans, climate-induced changes in ecosystems could have significant effects on human communities and economies.

The primary objective of this SAP is to provide a synthesis of the present state of scientific understanding to the climate and decision making communities on potential abrupt state changes or regime shifts in ecosystems in response to climate change. The Product is intended to be of value to:

- Policymakers in assessing current scientific capabilities to attribute causes of abrupt changes in ecosystems;

- Resource managers in developing alternative management strategies to enhance the resilience of ecosystems;

- Science program managers in identifying research needs and priorities that will enhance the ability to forecast and detect abrupt changes in ecosystems caused by climate.

This product is written primarily for the informed lay reader. For subject matter experts, more detailed discussions are available through the original references cited herein.

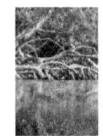

EXECUTIVE SUMMARY

Lead Authors: Daniel B. Fagre, USGS; Colleen W. Charles, USGS.
Authors: Craig D. Allen, USGS; Charles Birkeland, USGS-University of Hawaii; F. Stuart Chapin, III, University of Alaska; Peter M. Groffman, Institute of Ecosystem Studies; Glenn R. Guntenspergen, USGS; Alan K. Knapp, Colorado State University; A. David McGuire, USGS-University of Alaska; Patrick J. Mulholland, Oak Ridge National Laboratory; Debra P.C. Peters, USDA Agricultural Research Service; Daniel D. Roby, USGS-Oregon State University; George Sugihara, Scripps Institute of Oceanography and University of California at San Diego.
Contributing Authors: Brandon Bestelmeyer, Jornada Basin LTER; Julio L. Betancourt, USGS; Jeffrey E. Herrick, Jornada Basin LTER

ES.1. INTRODUCTION

As defined in this assessment, an ecological threshold is the point at which there is an abrupt change in an ecosystem quality, property, or phenomenon, or where small changes in one or more external conditions produce large and persistent responses in an ecosystem. Ecological thresholds occur when external factors, positive feedbacks, or nonlinear instabilities in a system cause changes to propagate in a domino-like fashion that are potentially irreversible. Once an ecological threshold is crossed, the ecosystem in question is not likely to return to its previous state.

Over the past three decades, climate change has become a recognized driver of ecosystem change. Changes in phenology, or shifts of species, and increases in such disturbances as wildland fires are all examples of ecosystem-scale responses to a warming biosphere. Much ecosystems research focuses on enhancing understanding of climate change impacts on ecosystems (and vice versa) and in developing the capability to predict the potential impacts of future climate change. In addition to the gradual types of climate-related change mentioned above, there is increasing recognition that small changes in climate can trigger major, abrupt responses in ecosystems when a threshold is crossed.

The potential for sudden, unanticipated shifts in ecosystem dynamics make resource planning, preparation, and management intensely difficult. These sudden changes to ecosystems and the goods and services they provide are not well understood, but they are extremely important if natural resource managers are to succeed in developing adaptation strategies in a changing world. This report provides an overview of what is known about ecological thresholds and where they are likely to occur. It also identifies those areas where research is most needed to improve knowledge and understand the uncertainties regarding them. The report suggests a suite of potential actions that land and resource managers could use to improve the likelihood of success for the resources they manage, even under conditions of incomplete understanding of what drives thresholds of change and when changes will occur. The focus of this report is on North American ecosystem threshold changes and what they mean for human society.

ES.2. EXAMPLES OF ECOSYSTEM THRESHOLDS

There are numerous examples of sudden ecological change that fit the current qualitative definition of an ecological threshold and that were likely caused by climatic changes such as warming temperatures. A clear example comes from recent observations of the Arctic tundra, where the effects of warmer temperatures have included reduced snow cover duration, which leads to reduced reflectivity of the surface. Reduced reflectivity causes greater absorption of solar energy, resulting in local warming, which, in turn, further accelerates the loss of snow cover. This amplified, positive feedback effect quickly leads to warmer conditions that foster the invasion of shrubs into the tundra. The new shrubs themselves then further reduce albedo and add to the local warming.

The net result is a relatively sudden, domino-like chain of events that result in conversion of the arctic tundra to shrubland, triggered by a relatively slight increase in temperature.

Examples like this illustrate the importance of positive feedbacks. Positive feedbacks are those that tend to increase alteration of the nature of the system, while negative feedbacks tend to minimize these changes. Ecosystems include both positive and negative feedbacks. Changes in external or internal factors that favor and strengthen positive feedbacks can lead to a change in conditions that may overwhelm other components of the system, leading to threshold changes. For example, the invasion and spread of a highly flammable grass in deserts will change the susceptibility of that landscape to fire. As another example, persistent drought will push an ecosystem's vegetation toward the limits of its physiological tolerance to water stress, creating conditions that favor drought-tolerant species at the expense of thirstier plants; this leads to system change, until a new state (with different, more drought-tolerant species) is achieved.

Ecosystems are not simple, and complex interactions between multiple factors and feedbacks can lead to even greater nonlinear changes in their dynamics. For example, the interaction of drought together with overgrazing can trigger desertification. Disturbance mechanisms, such as fire and insect outbreaks, shape many landscapes and may predispose many of them to threshold change when the additional stress of climate change is added. Furthermore, climate change will alter not only the landscape, but it will also affect the disturbance mechanisms themselves; in the example above, a warmer climate may not only lead to vegetation changes, but may also favor increased dryness, which will increase the likelihood of fire.

On a global scale, such altered disturbance regimes may influence rates of climate change. For example, as mentioned above, warm, dry conditions favor fire, and more fires release more carbon dioxide from burning vegetation, which in turn favors more warming. Adding additional complexity to already-complex systems, human actions also interact with natural drivers of change, producing multifaceted ecosystem changes that have important implications for the services provided by those ecosystems. For instance, the introduction of exotic, invasive plants may change the way in which an ecosystem responds to drought, and the conversion of woodland to farmed fields or urban areas will change the manner in which that landscape responds to intense storms.

The stories of several important ecosystems provide concrete examples of ecological thresholds, and illustrate the kinds of complex change that natural resource managers are facing, and that they must manage in the future.

As mentioned briefly above, a key example of observed climate-related threshold change is the warming of Alaska. Warming has caused a number of effects, including earlier snowmelt in the spring, reductions in sea-ice coverage, warming of permafrost, and resultant impacts to ecosystems including dramatic changes to wetlands, tundra, fisheries, and forests, including increases in the frequency and spatial extent of insect outbreaks and wildfire. During the 1990s, south-central Alaska experienced the largest outbreak of spruce bark beetles in the world. Milder winters and warmer temperatures increased the over-winter survival of the spruce bark beetle and allowed the bark beetle to complete its life cycle in 1 year instead of the normal 2 years. Added to this were 9 years of drought stress, which resulted in spruce trees that were too weak to fight off the beetle infestation. For these forests, multiple climate-triggered stresses amplified each others' effects to cause a profound ecosystem change.

The Alaskan spruce bark beetle outbreak and consequent forest die-off are an example of an actual climate-induced threshold crossing. There are additional ecosystems for which conditions suggest an approaching climate-related threshold. These include coral reefs, prairie pothole wetlands, and southwestern forests. Climate-related processes that affect coral reefs include sea-level rise, ocean acidification, and the increased water temperatures that are responsible for coral bleaching events. The Prairie Pothole Region of north-central North America is one of the most ecologically valuable freshwater resources of the Nation, with numerous wetlands that provide critical habitat for waterfowl populations. Climate models suggest a warmer, drier future climate for the Prairie

Pothole Region, which would result in a reduction in, or elimination of, wetlands that provide waterfowl breeding habitat. Similarly, predicted warmer, drier conditions in the semiarid forests and woodlands of the southwestern United States would place those forests under more frequent water stress, resulting in the potential for shifts between vegetation types and distributions, and could trigger rapid, extensive, and dramatic forest dieback.

In each of these cases, the anticipated changes would also be expected to tie to other nonlinear feedback relationships and other ecological disturbance processes, potentially leading to additional nonlinear threshold behaviors. Understanding and predicting the outcome of such complex interactions is not a trivial endeavor. Ecological systems are multivariate in nature, but current ecological forecasting model capabilities are comparatively simple and generally do not address the possibility or consequences of thresholds. Complex situations like those involving ecological thresholds thus tend to be beyond the limits of existing predictive capabilities. The end result is surprises for managers.

ES.3. RECOMMENDATIONS

If climate change is pushing more ecosystems toward thresholds, what can be done by land and resource managers and others to better cope with the threat of transformative change? Although the science of ecological thresholds is still in its infancy, one outcome of this synthesis effort was the identification of a suite of potential actions that, taken together or separately, can improve the understanding of thresholds and increase the likelihood of success in developing management and adaptation strategies in a changing climate, before, during, and after thresholds are crossed:

ES.3.1. Support Research To Identify Thresholds
While conceptually robust and widely acknowledged as already occurring, thresholds and threshold crossings have had relatively few empirical studies addressing them. Reliable identification of thresholds across different systems should be a national priority because of the potential for substantive surprises in the management of our natural resources.

ES.3.2. Enhance Adaptive Capacity
Given that threshold changes are increasingly likely to occur, it is important to prepare for them by increasing societal and ecological resilience. Managers that understand ecological diversity and the other factors that influence the resilience of the systems they manage are in a better position to implement changes that reduce the likelihood that thresholds will be crossed.

ES.3.3. Monitor and Adjust Multiple Factors and Drivers
Once the key factors controlling adaptive capacity and resilience are known, monitoring strategies should include those factors. Consideration should be given to monitoring indicators of ecosystem stress rather than the resources and ecological services of management interest.

ES.3.4. Develop Scenarios of the Consequences of Alternative Management Options for Dealing With Potential Changes
In some cases, the kinds of external factors that can precipitate threshold changes are well known, and furthermore are known in advance (for example, hurricanes, wildfire, or invasive species). In these cases, scenario analysis is a powerful tool for predicting and understanding the potential consequences of specific management actions.

ES.3.5. Collate and Integrate Information Better at Different Scales
Because agencies and institutions have different management mandates, there can be a focus on those resources and at their scales of interest to the exclusion of others. Better information sharing and integration have great potential for improving the understanding of thresholds and identifying when they might occur.

ES.3.6. Reduce Other Stressors
Many trigger points for abrupt change in ecosystems that are responding to climate change are not recognized, because human civilizations have not previously witnessed climate change of

this magnitude. However, other stressors for which reliable information exists can be reduced to make ecosystems healthier and more resilient as climate changes.

ES.3.7. Manage Threshold Shifts

There may be constraints to reducing or reversing climate-change-induced stresses to components of an ecosystem. If a threshold seems likely to occur but the uncertainties remain high as to when it will occur, contingency plans should be created. These can be implemented when the threshold shift begins to occur or can be carried out in advance if the approaching threshold is clear.

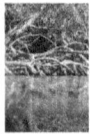

ES.3.8. Project Impacts to Natural Resources

There are many efforts to project climate change (for example, general circulation models) and ecosystem responses to climate change (for example, mapped atmosphere-plant-soil systems) using simulation modeling and other tools. These models generally project ecosystem trends and shifts, but do not explicitly consider the possibility of thresholds. A concerted effort must be made to understand, model, and project ecosystem responses to climate change with explicit acknowledgment of thresholds.

ES.3.9. Recognize Need for Decisionmaking at Multiple Scales

Much of the recent information on climate change impacts suggests that changes are occurring more quickly than forecast only a few years ago. It is also apparent that many changes are causing secondary, or cascading, domino-like changes in other parts of ecosystems. Management policies that were developed during relatively stable climate conditions may be inadequate for a variable world with more surprises. A shift toward multiple scales of information integration and subsequent decisionmaking can enhance and leverage existing management resources.

ES.3.10. Instigate Institutional Change To Increase Adaptive Capacity

In many cases, current institutional structures are geared towards disciplinary and jurisdictional isolation by agencies and, therefore, they do not facilitate synthesis across resources, regions, or issues. The capacity for such synthesis will be critical for identifying potential thresholds in ecosystem processes on multiple scales.

ES.3.11. Identify Research Needs and Priorities To Address Thresholds

At this point in time, very little is understood about thresholds in ecosystems. The major research needs and priorities that will enhance the ability in the future to forecast and detect abrupt changes in ecosystems caused by climate change must be articulated. The ubiquity of threshold problems across so many fields suggests the possibility of finding common principles at work. The cross-cutting nature of the problem of large-scale system change suggests an unusual opportunity to leverage effort from other fields and apply it to investigating systemic risk of crossing thresholds.

In summary, the science of understanding and predicting ecological thresholds is still in its infancy, and our existing understanding of many aspects and potential impacts of these thresholds is qualitative at best. The challenge is to improve the science needed to support decisionmaking, while recognizing that managing lands and resources is a continual process and that strategies are needed to inform management decisions that must be made under conditions of high uncertainties regarding potential thresholds. To better understand and prepare for ecological threshold crossings and their consequences, it is essential to increase the resilience of ecosystems and thus to slow or prevent the crossing of thresholds; to identify early warning signals of impending threshold changes; and to employ adaptive management strategies to deal with new conditions, new successional trajectories, and new combinations of species. Better integration of existing monitoring information across a range of spatial scales will be needed to detect potential thresholds, and research will need to focus on ecosystems undergoing a threshold shift to better understand the underlying processes. In a world being altered by climate change, natural resource managers may also have to be increasingly nimble, and adjust their goals for desired states of resources away from static, historic benchmarks and focus on increased resilience, biodiversity, and adaptive capacity as measures of success.

Introduction and Background

Lead Authors: Daniel B. Fagre, USGS; Colleen W. Charles, USGS.
Authors: Craig D. Allen, USGS; Charles Birkeland, USGS-University of Hawaii; F. Stuart Chapin, III, University of Alaska; Peter M. Groffman, Institute of Ecosystem Studies; Glenn R. Guntenspergen, USGS; Alan K. Knapp, Colorado State University; A. David McGuire, USGS-University of Alaska; Patrick J. Mulholland, Oak Ridge National Laboratory; Debra P.C. Peters, USDA Agricultural Research Service; Daniel D. Roby, USGS-Oregon State University; George Sugihara, Scripps Institute of Oceanography and University of California at San Diego.
Contributing Authors: Brandon Bestelmeyer, Jornada Basin LTER; Julio L. Betancourt, USGS; Jeffrey E. Herrick, Jornada Basin LTER

CHAPTER 1

1.1. THE PROBLEM OF SUDDEN CHANGE IN ECOLOGICAL SYSTEMS

The carbon dioxide (CO_2) concentration in the earth's atmosphere has reached 385 parts per million (ppm), a level that is unprecedented over the past one-half million years (based on ice core data) to 24 million years (based on soil data) (Hoegh-Guldberg *et al.*, 2007). CO_2 levels have been increasing during the past 150 years, with most of the change occurring in just the past few decades. Global mean temperature has risen in response to increased CO_2 concentration and is now higher than at any time in the past 1,000 years (based on tree rings) to 160,000 years [based on oxygen 18 (^{18}O) and deuterium (D) isotopes in ice]. The relatively sudden increase in the energy balance of the planet, due to an increase in greenhouse gas concentrations, has led to abrupt global climate changes that alter physical processes and biological systems on many scales and will certainly affect ecosystems that support human society (IPCC, 2007). One of the ways that a rapidly changing climate may affect ecosystems is by causing sudden, irreversible effects that fundamentally change the function and structure of the ecosystem with potentially huge impacts to human society (Wamelink *et al.*, 2003).

> Even small changes in physical conditions can provoke a regime shift that may not be easily or symmetrically reversed.

Even small, gradual change can induce threshold changes. For instance, in 1976–1977, major shifts occurred in sea surface temperatures, fisheries landings, zooplankton abundance, and community composition in the North Pacific (Hare and Mantua, 2000). Later analysis suggested that nonlinear regime shifts operate in this ecosystem, such that even small changes in physical conditions can provoke a regime shift that may not be easily or symmetrically reversed (for example, an increase in temperature from global warming, even as small as 0.5°C, has led to responses that have been well documented) (IPCC, 2007; Hsieh *et al.*, 2006). This tendency can be compounded by additional environmental stressors that predispose ecosystems to experience threshold changes in response to climate change. For example, in North America in the late 1990s, forests, woodlands, grasslands, and shrublands exhibited extensive dieback across the arid southwestern United States as overgrazing, fire suppression, and climate variability led to massive insect outbreaks and an unprecedented breadth of area consumed by fire (Allen, 2007).

Abrupt changes in ecosystems may result in dramatic reductions in ecosystem services, such as water supplies for human use. In the Klamath River basin in the Pacific Northwest, for example, the delicate socioecological balance of water allocation between needs for irrigated agriculture and habitat for endangered species of fish, which had been established in 1902, collapsed in 2002 during a multiyear drought because the system's resilience to maintain water quality in the face of climatic variability was degraded by long-term nutrient loading (NRC, 2002). Thresholds pose perhaps the greatest challenge currently facing climate change scientists. There is clear evidence that climate change has the potential to increase threshold changes in a wide range of ecosystems, but the basic and practical science necessary to predict and manage these changes is not well developed (Groffman *et al.*, 2006). In addition, climate change interacts with other natural processes to produce threshold changes. Disturbance mechanisms, such as fire and insect outbreaks (Crutzen and Goldammer, 1993; Lovett *et al.*, 2002, respectively), shape landscapes and may predispose many of them to threshold change when the stress of climate change is added (Swetnam and Betancourt, 1998). To complicate matters further, climate change can alter the disturbance mechanisms themselves, and on a global scale, altered disturbance regimes may influence rates of climate change. Another challenge is the multiscaled nature of threshold changes. These changes almost always involve coupled socioecological dynamics where human actions interact with natural drivers of change to produce complex changes in ecosystems that have important implications for the services provided by the ecosystems (Wamelink *et al.*, 2003).

A sense of urgency regarding thresholds exists because of the increasing pace of change, the changing features of the drivers that lead to thresholds, the increasing vulnerabilities of ecosystem services, and the challenges the existence of thresholds poses for natural resource management. These challenges include the potential for major disruption of ecosystem services and the possibility of social upheaval that might occur as new ways to manage and adapt for climate change and to cope with the unanticipated change are required.

Research on ecological thresholds is being assessed critically. The Heinz Center conducted several workshops that presented case studies of likely threshold change and began looking at possible social and policy responses. Another effort included numerous case studies focused on nonlinearities in ecological systems (Burkett *et al.*, 2005) and considered how thresholds are nonlinear responses to climate change. Recently, specific requests for proposals have been issued for research on thresholds (for example, see http://es.epa.gov/ncer/rfa/2004/2004_aqua_sys.html; http://cfpub.epa.gov/ncer_abstracts/index.cfm/fuseaction/reccipients.display/rfa_id/422/

records_per_page/ALL), and there are active efforts to bridge the gap between research and application in this area (see, for example, http://www.ecothresholds.org/). Assessment of the "state of the science" as it relates to ecosystems in the United States and for articulation of critical research needs is needed.

1.2. THE RESPONSE OF THE CLIMATE CHANGE COMMUNITY

Climate change is a very complex issue, and policymakers need an objective source of information about the causes of climate change, its potential environmental and socioeconomic consequences, and the adaptation and mitigation strategies to respond to the effects of climate change. In 1979, the first World Climate Conference was organized by the World Meteorological Organization. This conference expressed concern about man's activities on Earth and the potential to "cause significant extended regional and even global changes of climate" and called for "global cooperation to explore the possible future course of global climate and to take this new understanding into account in planning for the future development of human society" (IPCC, 2007). A subsequent conference in 1985 focused on the assessment of the role of CO_2 and other greenhouse gases in climate variations and associated impacts, concluding that an increase of global mean temperature could occur that would be greater than at any time in humanity's history. As a followup to this conference, the Advisory Group on Greenhouse Gases, a precursor to the Intergovernmental Panel on Climate Change (IPCC), was set up to ensure periodic assessments of the state of scientific knowledge on climate change and the implications of climate change for society. Recognizing the need for objective, balanced, and internationally coordinated scientific assessment of the understanding of the effects of increasing concentrations of greenhouse gases on the earth's climate and on ways in which these changes may potentially affect socioeconomic patterns, the World Meteorological Organization and the United Nations Environment Programme coordinated to establish an ad hoc intergovernmental mechanism to provide scientific assessments of climate change. Thus, in 1988, the IPCC was established to provide decisionmakers and others interested

Climate change can alter the disturbance mechanisms themselves, and on a global scale, altered disturbance regimes may influence rates of climate change.

in climate change with an objective source of information about climate change.

The role of the IPCC is to assess (on a comprehensive, objective, open, and transparent basis) the scientific, technical, and socioeconomic information relevant to understanding the scientific basis of risk of human-induced climate change, its potential impacts, and options for adaptation and mitigation and to provide reports on a periodic basis that reflect existing viewpoints within the scientific community. Because of the intergovernmental nature of the IPCC, the reports provide decisionmakers with policy-relevant information in a policy neutral way (IPCC, 2007). The first IPCC report was published in 1990, with subsequent reports published in 1995, 2003, and 2007.

In 1989, the U.S. Global Change Research Program began as a Presidential initiative and was codified by Congress in the Global Change Research Act of 1990 (Pub. L. 101–606), which mandates development of a coordinated interagency research program. The Climate Change Science Program (CCSP, http://www.climatescience.gov/), a consortium of Federal agencies that perform climate science, integrates the research activities of the U.S. Global Change Research Program with the U.S. Climate Change Research Initiative.

The CCSP integrates federally supported research on global change and climate change as conducted by the 13 U.S. Government departments and agencies involved in climate science. To provide an open and transparent process for assessing the state of scientific information relevant to understanding climate change, the CCSP established a synthesis and assessment program as part of its strategic plan. A primary objective of the CCSP is to provide the best science-based knowledge possible to support public discussion and government and private sector decisionmaking on the risks and opportunities associated with changes in the climate and related environmental systems.

The CCSP has identified an initial set of 21 synthesis and assessment products (SAPs) that address the highest priority research, observation, and decision-support needs to advance decisionmaking on climate change-related issues. This assessment, SAP 4.2, focuses on abrupt ecological responses to climate change,

or thresholds of ecological change. It examines the impacts to ecosystems when thresholds are crossed. It does not address those ecological changes that are caused by major disturbances, such as hurricanes. These externally driven changes, or exogenous triggers, are distinguished from changes caused by shifts in the ecosystem's response to a driver, such as a gradual rise in temperature. These internal changes in system response, or endogenous triggers, are the focus of this SAP. This SAP is one of four reports that address the Ecosystems research element and Goal 4 of the CCSP strategic plan to understand the sensitivity and adaptability of different natural and managed ecosystems and human systems to climate and related global changes.

I.3. THE GOAL OF SAP 4.2

This SAP summarizes the present state of scientific understanding regarding potential abrupt state changes or regime shifts in ecosystems in response to climate change. The goal is to identify specific difficulties or shortcomings in our current ability to identify the likelihood of abrupt state changes in ecosystems as a consequence of climate change.

Questions addressed by this SAP include:

1. What specifically is meant by abrupt state changes or regime shifts in the structure and function of ecosystems in response to climate change? What evidence is available from current ecological theory, ecological modeling studies, or the paleoecological record that abrupt changes in ecosystems are likely to occur in response to climate change?
2. Are some ecosystems more likely to exhibit abrupt state changes or threshold responses to climate change?
3. If abrupt changes are likely to occur in ecosystems in response to climate change, what does this imply about the ability of ecosystems to provide a continuing supply of ecosystem goods and services to meet the needs of humans?
4. If there is a high potential for abrupt or threshold-type changes in ecosystems in response to climate change, what changes must be made in existing management

A sense of urgency regarding thresholds exists because of the increasing pace of change, the changing features of the drivers that lead to thresholds, the increasing vulnerabilities of ecosystem services, and the challenges the existence of thresholds poses for natural resource management.

models, premises, and practices in order to manage these systems in a sustainable, resilient manner?

5. How can monitoring systems be designed and implemented, at various spatial scales, in order to detect and anticipate abrupt or threshold changes in ecosystems in response to future climate change?

6. What are the major research needs and priorities that will enhance the ability in the future to forecast and detect abrupt changes in ecosystems caused by climate change?

1.4. STANDARD TERMS

The 2007 Intergovernmental Panel on Climate Change Fourth Assessment Report (IPCC, 2007) is the most comprehensive and up-to-date report on the scientific assessment of climate change. This assessment (SAP 4.2) uses the standard terms defined in the IPCC's Fourth Assessment Report with respect to the treatment of uncertainty and the likelihood of an outcome or result based on expert judgment about the state of that knowledge. The definitions are shown in figure 1.1. This set of definitions is for descriptive purposes only and is not a quantitative approach from which probabilities relating to uncertainty can be derived.

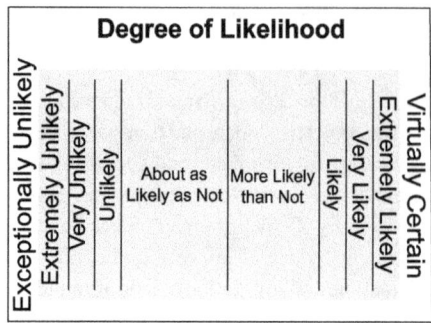

Likelihood terminology	Likelihood of the occurrence/outcome
Virtually certain	> 99% probability
Extremely likely	> 95% probability
Very likely	> 90% probability
Likely	> 66% probability
More likely than not	> 50% probability
About as likely as not	33 to 66% probability
Unlikely	< 33% probability
Very unlikely	< 10% probability
Extremely unlikely	< 5% probability
Exceptionally unlikely	< 1% probability

Figure 1.1. Degrees of outcome likelihood as defined in the IPCC's Fourth Assessment Report (AR4) (IPCC, 2007).

Ecological Thresholds

Lead Authors: Daniel B. Fagre, USGS; Colleen W. Charles, USGS.
Authors: Craig D. Allen, USGS; Charles Birkeland, USGS-University of Hawaii; F. Stuart Chapin, III, University of Alaska; Peter M. Groffman, Institute of Ecosystem Studies; Glenn R. Guntenspergen, USGS; Alan K. Knapp, Colorado State University; A. David McGuire, USGS-University of Alaska; Patrick J. Mulholland, Oak Ridge National Laboratory; Debra P.C. Peters, USDA Agricultural Research Service; Daniel D. Roby, USGS-Oregon State University; George Sugihara, Scripps Institute of Oceanography and University of California at San Diego.
Contributing Authors: Brandon Bestelmeyer, Jornada Basin LTER; Julio L. Betancourt, USGS; Jeffrey E. Herrick, Jornada Basin LTER

CHAPTER **2**

2.1. INTRODUCTION

Temperature, precipitation, and related climate variables are fundamental regulators of biological processes and it is reasonable to expect that significant changes in the climate system may alter linkages and feedbacks between ecosystems and regional climate systems. Increasing focus is being placed on the existence and likelihood of abrupt state changes or threshold responses in the structure and functioning of ecosystems (Holling, 1986; Scheffer *et al.*, 2001; Higgins *et al.*, 2002; Foley *et al.*, 2003; Schneider, 2004; Burkett *et al.*, 2005; Hsieh *et al.*, 2005). Various interrelated terms are employed in the scientific literature to characterize these types of discontinuous and rapid changes in ecosystems, including ecosystem tipping points, regime shifts, threshold responses, alternative or multiple stable states, and abrupt state changes. Our current understanding of thresholds and ecosystem responses makes it *unlikely* that we can predict such discontinuities in ecosystems, and these discontinuities are *likely* to result in profound changes to natural resources that are sensitive to climate changes, as well as to human societies that depend on ecosystem goods and services. This assessment, based on the literature and the synthesis teams' expertise, indicates that thresholds are *likely* to represent large-scale risk and uncertainty and will *likely* be a major challenge to natural resource managers.

On a predictive level, thresholds remain poorly understood, particularly in terms of the underlying causal mechanisms and the general factors that predispose systems to threshold effects.

Abrupt transitions have occurred in numerous ecosystems where incremental increases in global temperature have produced sudden and dramatic changes in the state of and the dynamics governing these systems (Anderson *et al.*, 2008). These thresholds of magnified ecological change are a consequence of the underlying nonlinear nature of ecosystems and are *very likely* critical to adaptation strategies for managing natural resources in a rapidly changing world. Sudden, unanticipated shifts in ecosystem dynamics are a major source of uncertainty for managers and make planning and preparation difficult. One of the primary objectives of this report (SAP 4.2) is to enhance the understanding and ability of managers to forecast the effects of climate change on ecosystems.

As discussed elsewhere in this chapter, the occurrence of threshold, or abrupt changes in ecosystems, is suggested by current ecological theory and models, and is documented with laboratory and field examples and even in the paleoecological record. However, on a predictive level, thresholds remain poorly understood, particularly in terms of the underlying causal mechanisms and the general factors that predispose systems to threshold effects. For example, it is unclear under what circumstances climate change (both in its mean state and in its variance in space and time, including occurrence of extreme weather events) might cause ecosystem threshold shifts, instead of more gradual, continuous changes in ecosystems and species. Further, it is not known what the resulting effects of very abrupt climate change (that is, crossing climate thresholds) on ecosystems will be. However, it will likely increase the likelihood of an ecosystem threshold shift. Thus, while rapid transitions in ecosystems are clear, reaching a level of understanding that enables one to anticipate or actually predict threshold effects is the main bottleneck to producing results that are useful to managers (Muradian, 2001; Bestelmeyer, 2006; Groffman *et al.*, 2006; Kinzig *et al.*, 2006).

2.2. EARLY DEVELOPMENT

The concepts of ecological thresholds, multiple stable states, and regime shifts originated in early theoretical work on the stability or persistence of ecosystems (Margalef, 1963; Lewontin, 1969; Odum, 1969; Holling, 1973; May, 1973, 1977). The two key components of stability were considered to be the system's "resilience," or the speed at which it would return to its current "stable equilibrium," and its "resistance," or ability to maintain its current "stable" state in the face of disturbance of a given magnitude. According to this early thinking, given enough disturbance, systems could be pushed into alternative stable states. This theoretical work was complemented (however sparsely) with early empirical demonstrations of multiple stable states in marine experimental systems (Sutherland, 1974) and with field data combined with model analysis for terrestrial ecosystems (Ludwig *et al.*, 1978).

"Stability" as a well-defined mathematical concept was central to these early theoretical discussions of thresholds. Lewontin (1969) reviewed mathematical models of stability and discussed the forces required to move an ecosystem out of a basin of attraction or stable state. May (1973) presented a precise definition of stability and a crater-and-ball analogy to illustrate the concepts. Later, May (1977) focused attention on the existence of alternative stable states and multiple equilibrium points with an emphasis on the thresholds between them. Holling (1973) drew attention to the ability of ecosystems to absorb and respond to disturbance and introduced the concept of resilience. Again, resilience focuses on dynamics far from equilibrium and was used to measure the magnitude of perturbations from which recovery of a system was no longer possible.

Although mathematically tractable and well defined in static engineering contexts, in the 1990s "stability" and the implication of "equilibrium" in ecological systems began gradually to give way to growing evidence that real ecological systems are neither static nor even well approximated as such. Notions of stable equilibrium, which continue to dominate much of our thinking and research to date are based on models and controlled experiments (for example, on paramecia and flour beetles) from the middle of the last century where singular static equilibrium was the ideal. Cracks in the equilibrium view began to appear as quantitative evidence mounted from natural systems demonstrating that "change" rather than "constancy" is the rule, and that nonlinear instability, thresholds, and chaos can be ubiquitous in nature (Dublin *et al.*, 1990; Sugihara and May, 1990; Tilman and Wedin, 1991; Grenfell, 1992; Knowlton, 1992; Hanski *et al.*, 1993; Sugihara, 1994). The possibility that so-called "pathological" nonequilibrium, nonlinear behaviors seen in theoretical treatments could be the rule in nature as opposed to a mathematical curiosity, opened the door for credible studies of thresholds. Indeed, threshold changes now appear to be everywhere. Recognition and documentation of sudden, not readily reversible changes in ecosystem structure and function have become a major research focus during the past 10 to 20 years (Scheffer *et al.*, 2001; Scheffer and Carpenter, 2003).

One of the important drivers of current interest in nonlinear ecosystem behavior and, in particular, threshold effects has been the recognition of the importance of unanticipated effects of climate change (Scholze *et al.*, 2006). Although much climate change research has focused on the direct effects of long-term changes in climate on the structure and function of ecosystems, there has been increasing recognition that the most dramatic consequences of climate change may occur as a result of indirect effects, including threshold changes (Vitousek, 1994; Carpenter, 2002; Schneider, 2004; Hobbs *et al.*, 2006).

2.3. CURRENT DISCUSSIONS OF THRESHOLD PHENOMENA

As ecologists were exploring the existence of alternative stable states in ecosystems, ocean-

There has been increasing recognition that the most dramatic consequences of climate change may occur as a result of indirect effects, including threshold changes.

ographers were documenting the impacts of major climatic events on the North Atlantic Ocean (Steele and Henderson, 1984), North Pacific Ocean, and Bering Sea ecosystems. They eventually used the term "regime shift" to describe the sudden shifts in biota that are driven by ocean climate events (Steele, 1996; Hare and Mantua, 2000). More recently, for the California Current Ecosystem (CCE), regime shifts in the biota have been distinguished from random excursions in the ocean climate based on the nonlinear signature of the time series (Hsieh *et al.*, 2006). The main idea here is that regimes represent different rules governing local dynamics (that is, they depend on environmental context), and that inherent positive feedbacks drive the system across thresholds into different dynamic domains. Thus, regime shifts in marine ecosystems are an amplified biological response to ocean climate variation (mainly temperature variation) rather than a simple tracking of environmental variation (Anderson *et al.*, 2008). On the other hand, ocean climate for the CCE in the 20th century did not have this nonlinear signature because the dynamic rules were the same in both warm and cold periods. Hsieh *et al.* (2006) and Anderson *et al.* (2008) suggest nonlinear forecasting methods as a rigorous way to detect thresholds because of the circularities of statistical methods. Current interest in regime shifts and thresholds in marine science has focused on understanding the factors that determine thresholds and on ways of extracting dynamics from observational data to make predictions.

Muradian (2001) and Walker and Meyers (2004) used a definition of regime shift developed by Scheffer and Carpenter (2003) emphasizing changes in the threshold level of a controlling variable in a system, such that the nature and extent of feedbacks change and result in a change in the system itself. Scheffer and Carpenter (2003) built on work in shallow lakes to demonstrate empirically the concept of threshold-like change and used these examples to further reinforce the idea that ecosystems are never stable but are dynamic and that fluctuations (in populations, environmental conditions, or ecosystems) are more the rule than not.

Given the move in thinking among many ecologists toward nonequilibrium and unstable

dynamics, the broader technical concept that may eventually replace "equilibrium" in this context is a more general notion concept that includes equilibrium, stable limit cycles, and nonequilibrium dynamics or chaos (Sugihara and May, 1990; Hsieh *et al.*, 2006). Depending on whether the control variable is thought of as part of the system (an intrinsic variable) or as external to the system (an extrinsic variable), threshold behavior may be thought of as a ridge of instability that separates control variables. From a more descriptive point of view, the idea suggests that there are particular states or characteristic combinations of species (grasslands, chaparral, oak-hickory forests, and so forth) that make up the biological component, and that ecosystem thresholds can be identified in the physical part of the system. Part of the nonlinearity or nonequilibrium nature of ecosystems comes from the fact that the biology (especially the dynamics) of the system is contingent on its own particular state (suite and abundance of species), as well as on the physical context in which it resides.

The field of range science has a parallel and largely independent literature on thresholds, resilience, regime shifts, and alternative stable states that has engendered a lively debate over how these terms are used in that field. Bestelmeyer (2006) argued that there is a lack of clarity in the use of the term "threshold" and its application to state-and-transition models that are used in range management. State-and-transition models describe alternative states and the nature of thresholds between states. Bestelmeyer's argument reflects a broad lack of consensus or understanding among range scientists about how best to define and use the threshold concept. Watson *et al.* (1996) criticized a focus on the consequences of threshold shifts at the expense of the processes that precede them. Many definitions of threshold phenomena emphasize relatively rapid, discontinuous phenomena (for example, Wissel, 1984; and Denoël and Ficetola, 2007). Others emphasize the points of instability at which systems collapse (Radford *et al.*, 2005) or the point at which even small changes in environmental conditions lead to large changes in state variables (Suding *et al.*, 2004). Still other definitions emphasize changes in controlling variables. According to Walker and Meyers (2004), "a regime shift involving alternative stable states occurs when a threshold level of a controlling variable in a system is passed."

There is clearly a need for these concepts to be tested across a wider set of ecosystems and have these experiments conducted with greater consistency and rigor to better evaluate the veracity of these concepts developed under rangeland conditions to other ecosystems and environmental conditions. One point of consensus underlying both the theoretical and empirical approaches to the topic of thresholds is that changes from one ecological condition to another take place around specific points or boundaries. But further advancement and agreement is limited by the small number of empirical studies that address this topic. While some believe that further advancement will depend on rigorous statistical testing for reliable identification of thresholds across different systems (Huggett, 2005), many in fields outside of range science see the danger of circularity in such arguments and suggest dynamic testing for determining threshold behavior (Hsieh *et al.*, 2005).

2.4. ECOLOGICAL THRESHOLDS DEFINED FOR SAP 4.2

Because of the variety of ways that the concept of thresholds has been developed, this assessment (SAP 4.2) uses the following general definition of ecological thresholds: *An ecological threshold is the point at which there is an abrupt change in an ecosystem quality, property, or phenomenon or where small changes in an environmental driver produce large, persistent responses in an ecosystem.* Fundamental to this definition is the idea that positive feedbacks or nonlinear instabilities drive the domino-like propagation of change that is potentially irreversible.

In line with this definition, threshold phenomena are particular nonlinear behaviors that involve a rapid shift from one ecosystem state (or dynamic regime) to another that is the result of (or that provokes) instability in any ecosystem quality, property, or phenomenon. Such instability always involves nonlinear amplification (some form of positive feedback) and is often the result of the particular structure of the interactions or the complex web of interactions. This definition distinguishes thresholds from other biological changes that are simple responses to external environmental change. Thus, bifurcation cascades (the point at which events take

one of two possible directions with important final consequences, making dynamic systems evolve in a nonlinear way with successive disruptions, divergences, or breaks from previous trends), nonlinear amplification (Dixon *et al.*, 1999), and the propagation of positive feedback (increasing instabilities) through complex webs of interactions are all interrelated attributes that fit our general working definition of threshold phenomena.

"Systemic" risk, or risk that affects the whole ecosystem rather than just isolated parts of the system, provides a useful analogy. Systemic risk corresponds to widespread change in an ecosystem characterized by a break from previous trends in the overall state of the system. Runaway changes are propagated by positive feedbacks (nonlinear instabilities) that are often hidden in the complex web of interconnected parts. Recovery may be much slower to achieve than the collapse, and the changes may be irreversible, in that the original state may not be fully recoverable (Chapin *et al.*, 1995). Our concept of threshold transitions include so-called bifurcation cascades where, for example, small changes in a controlling variable, such that the nature and extent of feedback change, leads to a sudden destabilization of the system.

Several examples of threshold crossings or transitions that illustrate this definition are described in Groffman *et al.* (2006). These include the interactions of drought and overgrazing that trigger runaway desertification, and the exceeding of some critical load, as with the toxicity limit of a contaminant or elimination of a keystone species by grazing, so that when one component of the system fails, it provokes a domino-like cascade of instability that substantially alters the rest of the system. Other examples are discussed in more detail in the case studies presented in chapter 3.

2.5. FACTORS THAT INFLUENCE RESILIENCE

At a general level, systems can be viewed as consisting of mixtures of positive and negative feedbacks, with positive feedbacks tending to alter the nature of the system, and negative feedbacks tending to minimize these changes (Chapin *et al.*, 1996). Changes that strengthen positive feedbacks (for example, the invasion and spread of highly flammable grass in a

An ecological threshold is the point at which there is an abrupt change in an ecosystem quality, property, or phenomenon or where small changes in an environmental driver produce large, persistent responses in an ecosystem.

desert) can lead to a change in conditions (for example, the fire regime) that may exceed the tolerance of other components of the system. This, in turn, leads to destabilization and threshold changes. Threshold crossings occur when positive feedbacks amplify changes in system characteristics in ways that exceed the buffering capacity of negative feedbacks that tend to maintain the system in its current state or the current limits of the control variables. Viewed from an adaptive management perspective, threshold crossings occur when changes in the system exceed the adaptive capacity of the system to adjust to change (Groffman et al., 2006). Because systems have adapted to the natural variability experienced in the past, anything that disrupts that variability can make them vulnerable to further change and amplified instability (Walker et al., 2006; Folke, 2006).

The following is a partial list of factors that are believed to come into play in determining a system's resilience, and sensitivity to threshold behavior (see also May and McLean, 2007):

1. A higher diversity of very weakly connected and substitutable components are thought to enhance resilience. Such arguments were made in the classic stability complexity debate [see reviews by Pimm (1984) and McCann (2000)].

2. Compartmentalization of interactions into guilds is a way to make model ecosystems more resilient to systemic events (May et al., 2008). Compartmentalization acts as a fire-break that prevents the spread of a system's collapse.

3. A predominance of weak linkages in a system with a few strong linkages leads to relatively low connectance (McCann, 2000; May et al., 2008) and is thought to increase resilience. Real ecological systems are thought to have a lognormal distribution of interaction strengths, which has been associated with increased resilience (Sala and Graham, 2002).

4. Ecosystems are resilient by virtue of their existence. They are the selected survivors of billions of years of upheaval and perturbation (continental drift, meteor extinctions, and so forth), and show some remarkable constancy in structure that persists for hundreds

of millions of years (for example, the constancy of predator-to-prey ratios). As such, enumerating the common attributes of these diverse naturally selected surviving systems, including those that change without experiencing thresholds, could be of interest to understanding thresholds.

5. Higher measured nonlinearity (greater instability) in the dynamics that provoke an increase in boom and bust population variability (Anderson et al., 2008) is directly associated with regime shifts. This is true in exploited marine fish populations, which show greater swings in abundance than their unexploited counterparts from the same environment. Exploited species show an amplified response to regime shifts, with greater extremes in abundance.

6. In line with the so-called "paradox of enrichment" (Rosenzweig, 1971), fertilizing a system to increase growth rates and carrying capacity can differentially advantage some species and provoke a rapid loss of species to a much simpler state.

7. Increasing time lags involved in population regulatory responses can destabilize systems (May, 1977), and this effect becomes more pronounced with higher growth rates. This is analogous to a large furnace (rapid growth) with a poor thermostat (regulatory delay), which tends to produce undershooting and overshooting of temperature in a way that predisposes the system to large-scale failure.

8. Reductions in variance, as might occur when managing systems for a stable flow of one particular good or service, tends to favor those species and components that are typical of this set of conditions at the expense of species that function more effectively under other conditions. Consequently the system as a whole remains stable under a narrower range of conditions.

2.6. THE BOTTOM LINE

To manage risks associated with ecological thresholds, it is essential to be able to forecast such events and to plan for and study alternative management scenarios. Because of the multi-scale nature of thresholds, better integration of existing monitoring information from the

> To manage risks associated with ecological thresholds, it is essential to be able to forecast such events and to plan for and study alternative management scenarios.

local to the largest possible spatial scales will be required to monitor and identify ecosystems that are approaching and undergoing critical transitions. Field research that focuses on ecosystems undergoing a threshold shift can help clarify the underlying processes at work. The rapid forest dieback in the southwestern United States, described in detail in the next chapter, is an example of a threshold shift for which field research identified the trigger (sudden tree mortality) that caused multiple other ecosystem changes. And natural resource managers will *very likely* have to adjust their goals for the desired states of resources away from historic benchmarks that may no longer be achievable in a nonequilibrium world that is continually changing and now being altered by climate change (Julius *et al.*, 2008). Such changes in methods and outlook as the following may be required:

- Abandon classic management models that assume a constant world in equilibrium (for example, maximum sustained yield models).

- Acknowledge in our management strategies and in our models that ecosystems are nonlinear, interdependent, and nonequilibrium systems.

- Use near-term forecasting tools, statistical and otherwise, that are appropriate to this class of system (for example, nonlinear time series prediction coupled with scenario models).

- Continue to identify the characteristics of systems that make them more or less vulnerable.

- Continue to identify early warning signals of impending threshold changes (and to monitor for those signals).

- Survey the major biomes to identify which systems might be most vulnerable to current climatic trends.

- Employ adaptive management strategies, such as skillful short-term forecasting methods coupled with scenario exploration models that are capable of dealing with new successional scenarios and novel combinations of species.

Case Studies

Lead Authors: Daniel B. Fagre, USGS; Colleen W. Charles, USGS.
Authors: Craig D. Allen, USGS; Charles Birkeland, USGS-University of Hawaii; F. Stuart Chapin, III, University of Alaska; Peter M. Groffman, Institute of Ecosystem Studies; Glenn R. Guntenspergen, USGS; Alan K. Knapp, Colorado State University; A. David McGuire, USGS-University of Alaska; Patrick J. Mulholland, Oak Ridge National Laboratory; Debra P.C. Peters, USDA Agricultural Research Service; Daniel D. Roby, USGS-Oregon State University; George Sugihara, Scripps Institute of Oceanography and University of California at San Diego.
Contributing Authors: Brandon Bestelmeyer, Jornada Basin LTER; Julio L. Betancourt, USGS; Jeffrey E. Herrick, Jornada Basin LTER

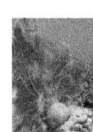

Thresholds of ecological change can occur at many spatio-temporal scales and in a diversity of ecosystems. The following examples were chosen to illustrate that thresholds probably have already been crossed in ecosystems in response to climate change and that the crossing of these thresholds will likely have implications at continental and global scales. Because these changes will likely impact American society significantly, these examples make clear the usefulness of considering thresholds in the monitoring and management of natural resources.

Four case studies are presented below in detail. They cover distinctly different types of ecosystems, all of which are potentially undergoing threshold-type changes. These studies are arranged in order of latitude, beginning with the highest. The first study is at a latitude in the far north where climate change has resulted in large temperature changes. The next study is of the mid-latitude Prairie Pothole Region where continental drying is expected because the subtropical high-pressure zone is broadening. The third case study is of forests of the West and Southwest, which are at slightly lower latitude, are generally already water-limited, and will be sensitive to the decreased water availability that will profoundly impact the western half of the United States. Finally, in the lowest latitude example, the effects of climate change in forcing threshold changes in coral reef ecosystems are examined.

3.1. CASE STUDY 1: ECOLOGICAL THRESHOLDS IN ALASKA

In recent decades, Alaska has warmed at more than twice the rate of the rest of the United States. The statewide annual average temperature has increased by 3.4°F since the mid-20th century, and the increase is much greater in winter (6.3°F). A substantial portion of the increase occurred during the regime shift of the Pacific Decadal Oscillation in 1976–1977. The higher temperatures of recent decades have been associated with changes in the physical environment, such as earlier snowmelt in the spring (Dye, 2002; Stone *et al.*, 2002; Dye and Tucker, 2003; Euskirchen *et al.*, 2006, 2007), a reduction of sea-ice coverage (Stroeve *et al.*, 2005), a retreat of many glaciers (Hinzman *et al.*, 2005), and a warming of permafrost (Osterkamp, 2007). In parallel with these changes in the physical environment, substantial changes in ecological systems have been observed, including major increases in the frequency of large-fire years in interior Alaska (Kasischke *et al.*, 2002), dramatic changes in the wetlands of interior Alaska (Yoshikawa and Hinzman, 2003), vegetation changes in the tundra of northern Alaska (Goetz *et al.*, 2005), and ecological changes that are affecting fisheries in the Bering Sea (Overland and Stabeno, 2004; Mueter and Litzow, 2008). Because Alaska is experiencing substantial changes in ecological systems, we divide the Alaska case study into four themes that focus on (1) changes in insect and wildfire regimes, (2) changes in wetlands, (3) vegetation change in northern Alaska, and (4) changes in Bering Sea Fisheries. For each of these themes we evaluate the occurrence and implications of threshold responses.

Ecological Thresholds and Changes in Insect and Wildfire Regimes of Interior Alaska.— Analyses of historical insect and fire disturbance in Alaska indicate that the extent and severity of these disturbances are intimately associated with longer and drier summers (Juday *et al.*, 2005; Balshi *et al.*, 2008). Between 1970 and 2000, the snow-free season increased by approximately 10 days across Alaska, primarily because of earlier snowmelt in the spring (Euskirchen *et al.*, 2006, 2007). Longer summers have the potential to be beneficial to the growth of plants; however, the satellite record suggests that the response of plant growth to warming differs in different regions of the State, with aboveground vegetation growth increasing in the tundra of northern Alaska and decreasing in the boreal forest of interior Alaska (Jia *et al.*, 2003; Goetz *et al.*, 2005). Analysis of forest growth data indicates that the growth of white spruce forests in interior Alaska is declining because of drought stress (Barber *et al.*, 2002), and there is the potential that continued warming could lead to forest dieback in interior Alaska (Juday *et al.*, 2005). The drought stress that has been experienced by trees in Alaska during recent decades makes them particularly vulnerable to attack by insects.

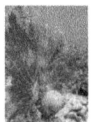

During the 1990s, south-central Alaska experienced the largest outbreak of spruce bark beetles in the world (Juday *et al.*, 2005). This outbreak was associated with a threshold response to milder winters and warmer temperatures that increased the overwinter survival of the spruce bark beetle and allowed the bark beetle to complete its life cycle in 1 year instead of the normal 2 years. This was superimposed on 9 years of drought stress between 1989 and 1997, which resulted in spruce trees that were too stressed to resist the infestation. The forests of interior Alaska are now threatened by an outbreak of spruce budworms, which generally erupt after hot, dry summers (Fleming and Volney, 1995). The spruce budworm has been a major insect pest in Canadian forests, where it has erupted approximately every 30 years (Kurz and Apps, 1999), but was not able to reproduce in interior Alaska before 1990 (Juday *et al.*, 2005). Areas that experience the death of trees over large areas of forest are vulnerable to wildfire, as the dead trees are highly flammable. This is of particular concern in interior Alaska where the frequency of large-fire years has been increasing in recent decades.

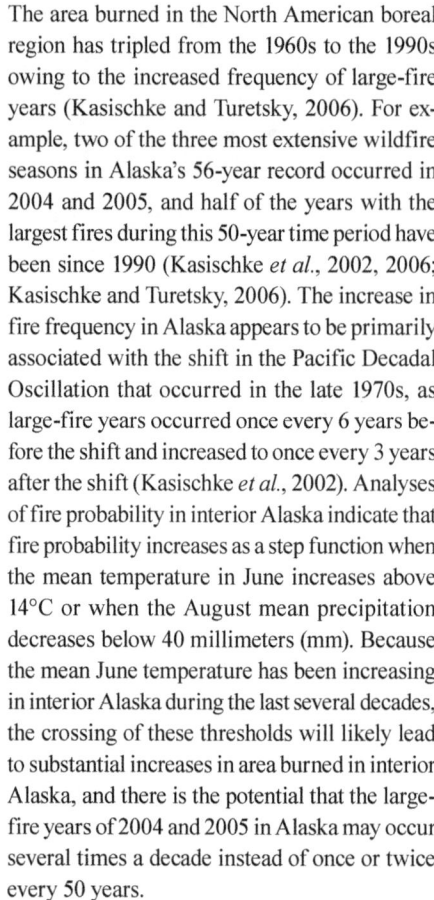

The area burned in the North American boreal region has tripled from the 1960s to the 1990s owing to the increased frequency of large-fire years (Kasischke and Turetsky, 2006). For example, two of the three most extensive wildfire seasons in Alaska's 56-year record occurred in 2004 and 2005, and half of the years with the largest fires during this 50-year time period have been since 1990 (Kasischke *et al.*, 2002, 2006; Kasischke and Turetsky, 2006). The increase in fire frequency in Alaska appears to be primarily associated with the shift in the Pacific Decadal Oscillation that occurred in the late 1970s, as large-fire years occurred once every 6 years before the shift and increased to once every 3 years after the shift (Kasischke *et al.*, 2002). Analyses of fire probability in interior Alaska indicate that fire probability increases as a step function when the mean temperature in June increases above 14°C or when the August mean precipitation decreases below 40 millimeters (mm). Because the mean June temperature has been increasing in interior Alaska during the last several decades, the crossing of these thresholds will likely lead to substantial increases in area burned in interior Alaska, and there is the potential that the large-fire years of 2004 and 2005 in Alaska may occur several times a decade instead of once or twice every 50 years.

Analyses of the response of fire to scenarios of future climate change indicate that the average area burned per year in Alaska will double by the middle of the 21st century for scenarios of both moderate and high rates of fossil fuel burning (Balshi *et al.*, 2008). By the end of the 21st century, fire is projected to triple in Alaska for a scenario of moderate rates of increase in fossil fuel burning and to quadruple for scenarios of high rates of increase in fossil fuel burning. Such increases have the potential to release large stocks of carbon stored in Alaska soils to the atmosphere, which would be a positive feedback to climate warming (Balshi *et al.*, 2008). The projected increase in the burned area also increases the fire risk to rural indigenous communities, reduces subsistence opportunities, and has implications for fire policy (Chapin *et al.*, 2008).

Ecological Thresholds and Changes in Wetlands of Interior Alaska.—There has been a documented decrease in the area of closed-basin lakes (that is, lakes without stream inputs and outputs) during the latter half of the 20th century in the southern two-thirds of Alaska (Klein *et al.*, 2005; Riordan *et al.*, 2006). The decrease in lake area appears to be caused by greater evaporation associated with longer and drier summers and by sudden drainage associated with thawing of permafrost in areas where the temperature of permafrost is close to melting. A decrease in the area of closed-basin lakes has also been documented in Siberia in areas of "warm" permafrost (Smith *et al.*, 2005).

Discontinuous permafrost in Alaska is warming and thawing, and extensive areas of thermokarst terrain (marked subsidence of the surface resulting from thawing of ice-rich permafrost) are now developing as a result of climatic change. Estimates of the magnitude of the warming at the discontinuous permafrost surface are 0.5° to 1.5°C (Osterkamp and Romanovsky, 1999). Thermokarst is developing in the boreal forests of Alaska where ice-rich discontinuous permafrost is thawing. Thaw subsidence at the thermokarst sites is typically 1 to 2 meters (m) with some sites experiencing subsidence of up to 6 m (Osterkamp *et al.*, 1997). Much of the discontinuous permafrost in Alaska is warm and is highly susceptible to thermal degradation if regional warming continues. Warming of permafrost may be causing a significant loss of open water across Alaska as thawing of permafrost connects closed watersheds with groundwater (Yoshikawa and Hinzman, 2003).

Examination of satellite imagery indicates that the loss of water can occur suddenly, which suggests catastrophic drainage associated with thawing of permafrost (Riordan *et al.*, 2006). However, the reduction of open water bodies may also reflect increased evaporation under a warmer and effectively drier climate as the loss of open water has also been observed in permafrost-free areas (Klein *et al.*, 2005).

In wetland complexes underlain by ice-rich permafrost in areas of hydrologic upwelling (for example, wetland complexes abutting up against the foothills of large mountain ranges), the thawing of that permafrost may result in wetland expansion as trees die when their roots are regularly flooded, causing wet sedge meadows, bogs, and thermokarst ponds and lakes to replace forests (Osterkamp *et al.*, 2000). The Tanana flats, which extends nearly 70 miles from the northern foothills of the Alaska Range to Fairbanks, Alaska, is underlain by ice-rich permafrost that is thawing rapidly and causing birch forests to be converted

to minerotrophic floating mat fens (Jorgenson *et al.*, 2001). It is estimated that 84 percent of a 260,000-hectare (ha) (642,000-acre) area of the Tanana flats was underlain by permafrost a century or more ago. About one-half of this permafrost has partially or totally degraded. These new ecosystems favor aquatic birds and mammals, whereas the previous forest ecosystems favored land-based birds and mammals.

During the past 50 years, it appears that warming has generally resulted in the loss of open water in closed-basin lakes in wetland complexes located in areas of discontinuous permafrost in the southern two-thirds of Alaska (Riordan *et al.*, 2006). The Tanana flats near Fairbanks is the only area where an increase in water area has been documented (Jorgenson *et al.*, 2001), and closed-basin lakes in the tundra region of northern Alaska have shown no changes in area during the past 50 years (Riordan *et al.*, 2006). The loss of area of closed-basin lakes in interior Alaska may be indicative of a lowering of the water table that has the potential to convert wetland ecosystems in interior Alaska into upland vegetation. A substantial loss of wetlands in Alaska has profound consequences for management of natural resources on national wildlife refuges in Alaska, which cover about 3.1 million hectares (more than 77 million acres) and make up 81 percent of the National Wildlife Refuge System. These refuges provide breeding habitat for millions of waterfowl and shorebirds that winter in more southerly regions of North America. Reduction of habitat area would present a substantial challenge for waterfowl management across the National Wildlife Refuge System (Julius *et al.*, 2008). Wetland areas have also been traditionally important in the subsistence lifestyles of native peoples in interior Alaska, as many villages are located adjacent to wetland complexes that support an abundance of wildlife subsistence resources. Thus, the loss of wetland area has the potential to affect the sustainability of subsistence lifestyles of indigenous peoples in interior Alaska.

Ecological Thresholds and Vegetation Changes in Northern Alaska.— Shrub cover in northern Alaska has increased by about 16 percent since 1950 (Sturm *et al.*, 2001; Tape *et al.*, 2006), and the tree line in Alaska is expanding in most places (Lloyd and Fastie, 2003; Lloyd, in press). This is consistent with satellite observations, which show an approximately 16 percent increase per decade in the normalized difference vegetation index (NDVI) (Jia *et al.*, 2003; Goetz

Discontinuous permafrost in Alaska is warming and thawing, and extensive areas of thermokarst terrain are now developing as a result of climatic change.

et al., 2005). The increased growth of vegetation at or above the tree line appears to be a response to longer and warmer growing seasons. Tundra vegetation in northern Alaska may not be experiencing drought stress to the extent experienced by forests in interior Alaska because the surface water in tundra regions is not able to drain away through the ice-rich continuous permafrost. Experimental studies demonstrate that arctic summer warming of 1°C increases shrub growth within a decade (Arft *et al.*, 1999). Satellite analyses of relationships between NDVI and summer warming (Jia *et al.*, 2003) suggest that the response of tundra vegetation is linearly related to summer warmth. Thus, it appears that the response of tundra vegetation to warming is not a threshold response.

While growth of shrubs and trees may not be threshold responses to warming, the changing snow cover and vegetation in northern Alaska have the potential to result in sudden changes in the absorption of heat from incoming solar radiation and the transfer of that heat to warm the atmosphere. For example, the advance in snowmelt reduces spring albedo, causing the ecosystem to absorb more heat and transfer it to the atmosphere. The snowmelt-induced increase in heating in northern Alaska has been about 3.3 watts per square meter (W m^{-2}) averaged over the summer, similar in magnitude to the 4.4 W m^{-2} caused by a doubling of atmospheric CO_2 over several decades (Chapin *et al.*, 2005). Thus, gradual warming has caused a rapid advance in the snowmelt date and a very large increase in local heating. Although vegetation changes to date have had minimal effects on atmospheric heating, conversion to shrubland would increase summer heating by 8.9 W m^{-2}, with even larger changes triggered by conversion to forest. Warming experiments that increase shrubs also reduce the abundance of lichens, an important winter food of caribou (Cornelissen *et al.*, 2001). Most arctic caribou herds are currently declining in population, although the reasons are uncertain. In summary, positive feedback associated with earlier snowmelt and shrub expansion is amplifying arctic warming and may be altering food-web dynamics in ways that have important cultural and nutritional implications for northern indigenous people.

Ecological Thresholds and Fisheries of the Bering Sea.—Alaska leads the United States in the value of its commercial fishing catch, and most of the Nation's salmon, crab, and herring come from Alaska, and specifically from the Bering Sea. The Bering Sea is one of the most productive marine ecosystems in the world, supporting some of the largest oceanic populations of fish, seabirds, and marine mammals anywhere (Loughlin *et al.*, 1999). The Bering Sea provides 47 percent of total U.S. fishery production by mass, including the largest single species fishery in the United States, walleye pollock (*Theragra chalcogramma*) (Criddle *et al.*, 1998). It is also an important source of subsistence resources (such as, fish, marine mammals, and seabirds) for more than 30 Alaska Native communities and supports 95 percent of the worldwide population of northern fur seals, 80 percent of the total number of seabirds that breed in the United States, and major populations of tens of thousands of Pacific walrus, Steller sea lion, and several species of great whales. This production is fueled by nutrients annually replenished from slope and oceanic waters across the very broad (more-than-500-kilometer-wide) continental shelf (Stabeno and Overland, 2001; Stabeno *et al.*, 2006).

Changes in fisheries of the Bering Sea occurred during and after the transition from cool to warm conditions in 1976–1977, in association with a regime shift in the Pacific Decadal Oscillation, and were followed by historically high commercial catches of salmon and pollock, as well as a shift away from crab dominance on the ocean floor (Overland and Stabeno, 2004). In the past decade, geographic displacement of marine mammal populations to the north has been documented in the Bering Sea region (Moore *et al.*, 2003). The displacements of walrus and seal populations are already apparent to coastal communities. The northward displacements of fauna in the Bering Sea has coincided with a reduction of benthic (organisms that live on or near the ocean floor) prey populations, an increase and northward shift in pelagic (those of the open seas and oceans) fishes, an increase in air and ocean temperatures, and a reduction in sea ice (Stroeve *et al.*, 2005; Grebmeier *et al.*, 2006).

Ultimately, populations of fish, seabirds, seals, walruses, and other species depend on water temperatures and plankton blooms that are regulated by the thickness, extent, and location of the ice edge in spring (Hunt and Stabeno, 2002). As the sea ice continues to retreat, the location, timing, and species makeup of the blooms is changing, subarctic pelagic food webs are replacing arctic ones, and the amount of food reaching the living

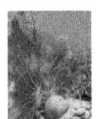

The changing snow cover and vegetation in northern Alaska have the potential to result in sudden changes in the absorption of heat from incoming solar radiation and the transfer of that heat to warm the atmosphere.

things on the ocean floor, the benthos, is declining dramatically. This in turn radically changes the species makeup and populations of fish and other marine life forms, with significant repercussions for fisheries (Anderson and Piatt, 1999; Litzow *et al.*, 2008; Hatfield *et al.*, 2008; Julius *et al.*, 2008). Reductions in sea-ice cover also result in reduced albedo (reflectance of solar radiation), greater sea surface temperatures, and accelerated sea-ice retreat, a positive feedback loop that is at least partly responsible for the unexpected and record-setting extent of open water in the Arctic Ocean in recent years. Thus, changes in sea ice are the major driver of concern with respect to threshold changes in fisheries of the Bering Sea (Mueter and Litzow, 2008).

Seasonal sea-ice extent currently divides the Bering Sea eastern shelf into two biogeographic provinces, which differ in production pathways. In the subarctic biogeographic province (south of the average annual maximum extent of the sea ice), most primary production remains within the pelagic ecosystem, and pollock is the dominant tertiary consumer (Macklin and Hunt, 2004). In contrast, in the arctic biogeographic province, tight coupling between pelagic primary production and the benthos benefits benthic foragers, such as gray whales, walrus, and diving ducks (Lovvorn *et al.*, 2003; Grebmeier *et al.*, 2006). The boundary between the two biogeographic provinces varies in location on longer time scales (decadal or longer) and is expected to move northward as the region becomes warmer. The average southern edge of the maximum ice extent currently lies north of the Pribilof Islands (Byrd *et al.*, 2008).

The Bering Sea ecosystem, however, is in a state of rapid flux due to climate change. Present data and climate projections from atmosphere-ocean models predict major loss of sea ice during the next few decades (Overland and Stabeno, 2004; Holland *et al.*, 2006); the Bering Sea is particularly sensitive to global warming because of the seasonal nature of sea-ice cover (Grebmeier *et al.*, 2006). Recent relative temperature extremes (above 2°C) in Alaska and adjacent waters represent the largest recent change on the planet (Hansen *et al.*, 2006). However, these models and empirical data also demonstrate large natural variability. Ecosystems will *likely* be affected by how the path of such warming occurs—that is, whether there will be a continued slow warming trend

with little interannual variability (in which case crossing of ecological thresholds is less likely) versus a warming trend that incorporates wide swings in temperature and extent of sea ice (enhancing the likelihood of threshold crossings). Climatic and oceanographic conditions in the Bering Sea during 2007–2008 were unexpectedly cold, supporting the latter scenario.

Warming of the Bering Sea is altering the geographic distributions and behaviors of humans, marine mammals, seabirds, and fish by restructuring their habitats and food webs (Grebmeier *et al.*, 2006; Mueter and Litzow, 2008). As a result of warming, changes in the time and place of food production lead to dominance of top-down control processes in the pelagic marine environment and the decline of benthic production. Under a long-term warming scenario with early ice retreat, bottom-up control mechanisms (temperature, sea-ice extent and duration, ocean currents, and nutrient fluxes) set the stage for the emergence and dominance of top-down control processes in the pelagic marine environment and the decline of benthic production (Mueter and Litzow, 2008), a threshold change akin to that was documented after the 1976–1977 regime shift in the Pacific Decadal Oscillation. Increased heat content would increase the combined populations of the subarctic piscivores—arrowtooth flounder, pollock, and cod—in proportion to expanded breeding grounds and increased availability of food during critical developmental stages (Hunt and Stabeno, 2002). Because arrowtooth flounder is not targeted by fishing, it is likely to become the dominant component of the biomass of the three subarctic piscivores in this system and is predicted to be one of the principal agents of top-down control in the Bering Sea, as predator and competitor of the now-dominant, but commercially exploited, pollock and cod. Such a rapid and dramatic restructuring of subarctic marine communities is not unprecedented; the 1976–1977 regime shift in the Pacific Decadal Oscillation resulted in threshold community reorganization in the Gulf of Alaska (Anderson and Piatt, 1999).

Arrowtooth flounder is also an agent of change as a direct and indirect competitor of fur seals, murres, kittiwakes, and other top trophic-level piscivores for their respective forage species (juvenile pollock, capelin, sand lance, herring, and myctophids). Populations of fur seals, murres, and kittiwakes could decline or increase in the

near term, depending on the locality of rookeries and nesting colonies, but long-term overall trends would be downward under warming. Fur seals, murres, and kittiwakes would further decline owing to competition from humpback and fin whales, with fur seal declines being further accelerated by increasing killer whale predation. Dislocation of feeding hot spots would likely disadvantage breeding fur seals, murres, and kittiwakes as central place foragers, but would work to the advantage of humpback and fin whales, further exacerbating direct and indirect competition between these two groups of species. Dislocations and declines in fur seals, kittiwakes, murres, pollock, and cod would stress human communities by increasing the costs of maintaining a livelihood and obtaining food and by necessitating changes in the types of food taken and the means of harvest. Both commercial fishers based in Dutch harbor and subsistence fishers based in over 30 Native Alaskan communities on the shores of the Bering Sea are facing greater commuting distances and higher risks to exploit fisheries resources that were formerly close to home.

The northern Bering Sea, in particular, is experiencing a rapid shift in the structure and function of the formerly arctic community to conditions typical of marine ecosystems of the subarctic (Hunt *et al.*, 2002; Grebmeier *et al.*, 2006). The earlier sea-ice retreat results in a later, warm-water spring phytoplankton bloom, increased grazing by zooplankton, and greater pelagic secondary productivity (Hunt *et al.*, 2002). Concurrently, benthic productivity is decreasing (Grebmeier *et al.*, 2006). The formerly ice-dominated, shallow marine ecosystem that favored highly productive benthic communities also supported high densities of upper trophic-level bottom-feeders, such as Pacific walruses, gray whales, and seaducks, including the Endangered Species Act (ESA)-listed spectacled eider.

The northward flowing Anadyr Current, which originates in the southern Bering Sea, transports nutrient-rich water far onto the Bering Shelf and the northern Bering Sea. This largely wind-forced transport creates highly productive shelf waters in the area north of St. Lawrence Island and south of the Bering Strait, known as the Chirikov Basin (Springer *et al.*, 1989; Piatt and Springer, 2003). Oceanic copepods, such as *Neocalanus cristatus* and *N. flemingeri*, transported by the Anadyr Current, along with the large euphausiid *Thysanoessa raschii* provide abundant prey for planktivores

foraging near St. Lawrence Island (Piatt *et al.*, 1988). The Anadyr Current is highly variable on a seasonal and annual basis, usually reaching its greatest velocity during July (about 1.3 Sv, or 13 million cubic meters per second) (Roach *et al.*, 1995). Consequently, the primary productivity on the Bering Shelf during summer months varies with the strength of northward flow associated with the Anadyr Current (Springer *et al.*, 1989; Russell *et al.*, 1999).

When the Anadyr Current is weaker, planktivores presumably rely more on zooplankton associated with northern Bering Shelf waters, such as the small copepod *Calanus marshallae* and the large amphipod *Themisto libellula* (Coyle, Chavtur, and Pinchuk, 1996; Russell *et al.*, 1999). *Neocalanus* copepods are larger and have higher energy content per prey item than the small, neritic copepod *C. marshallae*, which is characteristic of Bering Shelf water. The lipid content of *Neocalanus* copepods is also probably higher (Obst *et al.*, 1995), making these oceanic species more energy-dense than their shelf domain counterparts. When preferred *Neocalanus* copepods are not available, planktivores must switch to other prey types. The progressively earlier transition from winter to spring in the Bering Sea, changes in prevailing weather patterns and associated wind forcing, and the resulting changes in primary and secondary productivity are expected to have large impacts on upper trophic-level consumers that rely on the Anadyr Current (Stabeno and Overland, 2001; Grebmeier *et al.*, 2006).

Projected warming of the Bering Sea is also expected to profoundly alter the structure of the southeastern Bering Sea ecosystem by changing pathways and fluxes of energy flow, as well as the abundance, spatial distribution, and species composition of fish, seabirds, and marine mammals, thereby affecting commercial and subsistence fisheries that support local, regional, and national economies (Hunt and Stabeno, 2002; Grebmeier *et al.*, 2006; Mueter and Litzow, 2008). Climate-induced changes in physical forcing of the Bering Sea modify the partitioning of food resources at all trophic levels on the continental shelf through bottom-up processes. The emergent properties of this formerly seasonal sea-ice-dominated marine ecosystem under warming are still the subject of intense scientific inquiry, but the weight of evidence suggests that the Bering Sea ecosystem has reached a threshold of major ecosystem change and community restructuring.

3.2.   CASE STUDY 2. THE MID-CONTINENT PRAIRIE POTHOLE REGION: THRESHOLD RESPONSES TO CLIMATE CHANGE

The Prairie Pothole Region of north-central North America is one of the most ecologically valuable freshwater resources of the Nation (van der Valk, 1989). It contains 5 million to 8 million wetlands, which supply critical habitat for continental waterfowl populations and provide numerous valuable ecosystem services for the region and Nation. The weather extremes associated with this region are particularly important for the long-term productivity of waterfowl dependent on these wetlands.

The Prairie Pothole Region (fig. 3.1) exhibits a variable climate, ranging from severe droughts, exemplified by the 1930s when agriculture was devastated, grassland communities shifted eastward, and trees died by the millions (Albertson and Weaver, 1942, 1945; Woodhouse and Overpeck, 1998; Rosenzweig and Hillel, 1993), to periods of deluge, such as occurred in the late 1900s when closed-basin lakes flooded, causing high mortality of shoreline trees and considerable economic damage to farmland, roads, and towns (Winter and Rosenberry, 1998; Johnson

et al., 2005; Shapley *et al.*, 2005). The 20th-century climate of the Prairie Pothole Region was punctuated by significant droughts. These conditions have occurred over small and large areas and lasted as short as several growing seasons to as long as a decade (Skaggs, 1975; Laird and Cumming, 1998; Nkemdirim and Weber, 1999).

Wetlands in the Prairie Pothole Region are likely to be strongly affected by gradual changes in climate (Poiani and Johnson, 1991; Covich *et al.*, 1997). Climate drives surface processes, such as the hydrologic cycle, and hydrology is the most important factor that controls key wetland processes and services (Winter and Woo, 1990). A warmer and drier climate, as indicated by general circulation models for the northern Great Plains (Ojima and Lackett, 2002), could affect the wetland hydroperiod, the ratio of emergent plant cover to open water, the species composition, wetland permanence, and primary and secondary productivity, among others (van der Valk, 1989). Winter (2000) predicted that

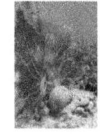

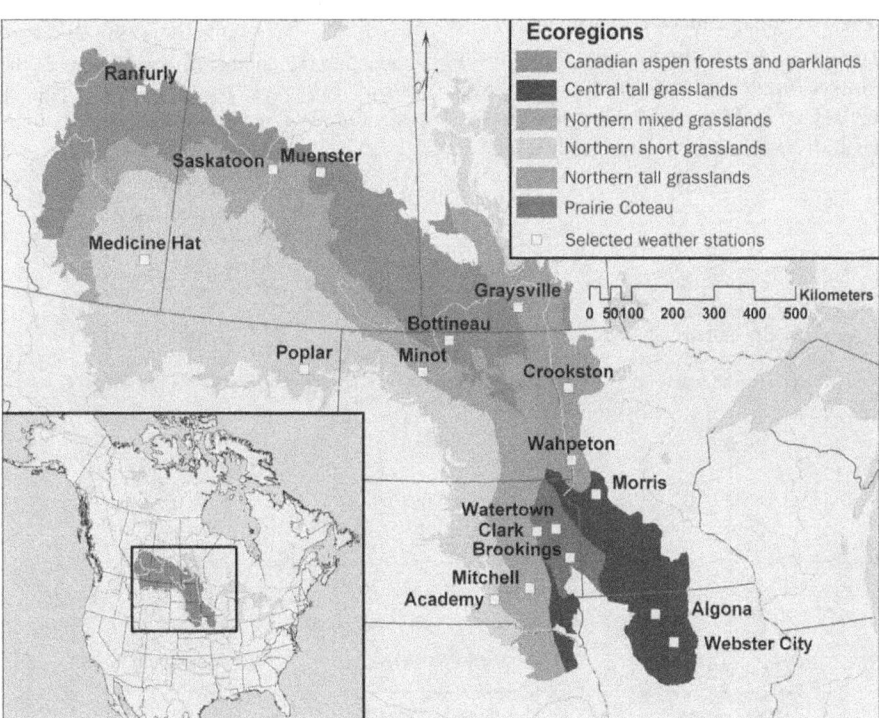

Figure 3.1. Location of the Prairie Pothole Region (PPR) of North America (red highlighted area). (Johnson *et al.*, 2005)

the surface area of seasonal and semipermanent wetlands in the Prairie Pothole Region would be reduced by increases in evapotranspiration and reduced summer soil moisture. With increased temperatures, summer evapotranspiration would put increasing demands on groundwater, resulting in earlier drying of wetlands. Thus, additional climate variability of the magnitude suggested by global climate change models would profoundly affect wetland water budgets and the many processes and attributes linked to these wetlands.

Changing climate can have direct effects on the trajectories of these wetland ecosystems and their sustainability. Shifts in climate in this region over decadal time scales could result in longer or more frequent drought periods and may lead to threshold responses by the wetland systems. The interaction of extrinsic and intrinsic processes reflected in such hydrologically, geologically, and biologically linked systems as wetlands and their surrounding watersheds could result in rapid nonlinear changes at broad spatial scales that are triggered by small differences in temperature and precipitation if threshold values are exceeded that may also result in these systems exhibiting hysteresis.

The first quantitative assessments of the possible effects of climate change on Prairie Pothole Region wetlands used the WETSIM (WETland SIMulator), which is a rule-based, spatially explicit simulation model that is composed of hydrology and vegetation sub-models (Poiani and Johnson, 1991, 1993a, b; Poiani *et al.*, 1995, 1996). Simulations using this model and general circulation model climate forcings indicate that semipermanent wetlands would lose their historic highly dynamic character by drying up more frequently and becoming chronically choked with emergent cover. Shortened hydroperiods and continuous stands of emergent cover for semipermanent wetlands across the Prairie Pothole Region would have strong negative effects on the continental population of water birds (particularly ducks).

Johnson *et al.* (2005) used a simulation model (WETSIM) to contrast historical and future wetland conditions across the Prairie Pothole Region of North America (fig. 3.1). They assembled 95-year climate data sets for 18 weather stations across the Prairie Pothole Region as input to a revised version of WETSIM (version 3.1), which enabled a much broader geographic assessment to be conducted of the effects of past and future climate variability on wetland conditions across the Prairie Pothole Region. Their model runs reflected the high level of spatial and temporal heterogeneity in wetland water levels historically across the Prairie Pothole Region. They were able to use model output to simulate the number of completions of the wetland cover cycle across the Prairie Pothole Region (fig. 3.2; Weller, 1965).

The interaction of extrinsic and intrinsic processes reflected in such hydrologically, geologically, and biologically linked systems as wetlands and their surrounding watersheds could result in rapid nonlinear changes at broad spatial scales that are triggered by small differences in temperature and precipitation.

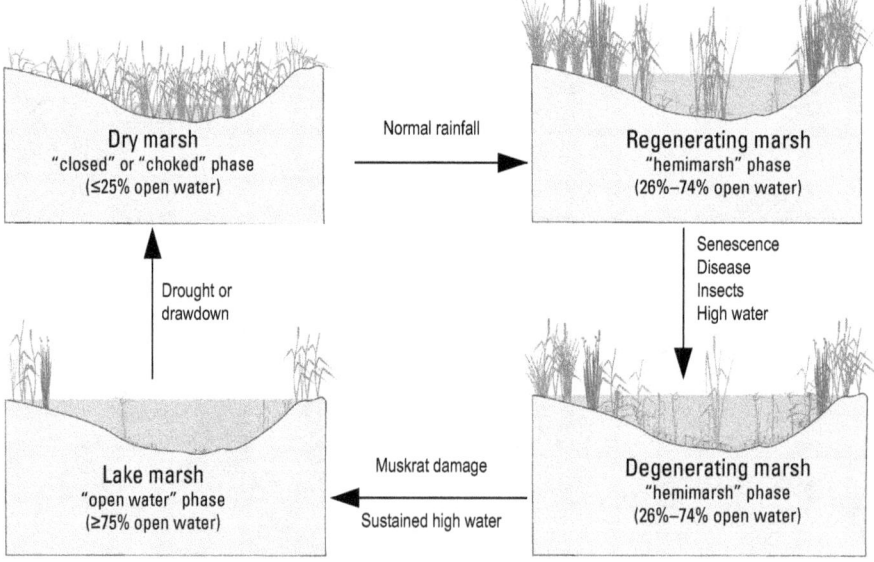

Figure 3.2. Wetland cover cycle (modified from Weller, 1965).

The wetland cover cycle was highly sensitive to alternative future climates. The geographic pattern of return times shifted markedly with changes in temperature and precipitation. A 3°C increase in temperature and no change in precipitation resulted in a greatly diminished area and geographic shift eastward for the region of fastest return times. However, reduced precipitation and warmer air temperatures resulted in no complete cover cycle return times across the Prairie Pothole Region except in a small area of north-central Iowa (fig. 3.3), thus representing a threshold response to climate change. Such dramatic shifts in wetland conditions emphasize the sensitivity of Prairie Pothole Region wetlands to climate variability.

Using this information, Johnson *et al.* (2005) simulated the occurrence of highly favorable water and cover conditions for breeding waterfowl (fig. 3.4). The most productive habitat for breeding water birds would shift under an effectively drier climate from the center of the Prairie Pothole Region (the Dakotas and southeastern Saskatchewan) to the wetter eastern and northern fringes (in sync with the changes in the cover cycle return results).

Continental waterfowl population cycles are largely dictated by regional wetland conditions, with population declines being commonplace during periods of drought and then rebounding during wetter periods. Under a warmer, drier climate, wetlands would be especially vulnerable even if precipitation were to continue at historic levels (Johnson *et al.*, 2005). The geographic shifts in the most favorable region for waterfowl breeding resulting from the model simulation runs will *likely* affect the rate at which the threshold for waterfowl population sustainability will be reached.

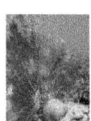

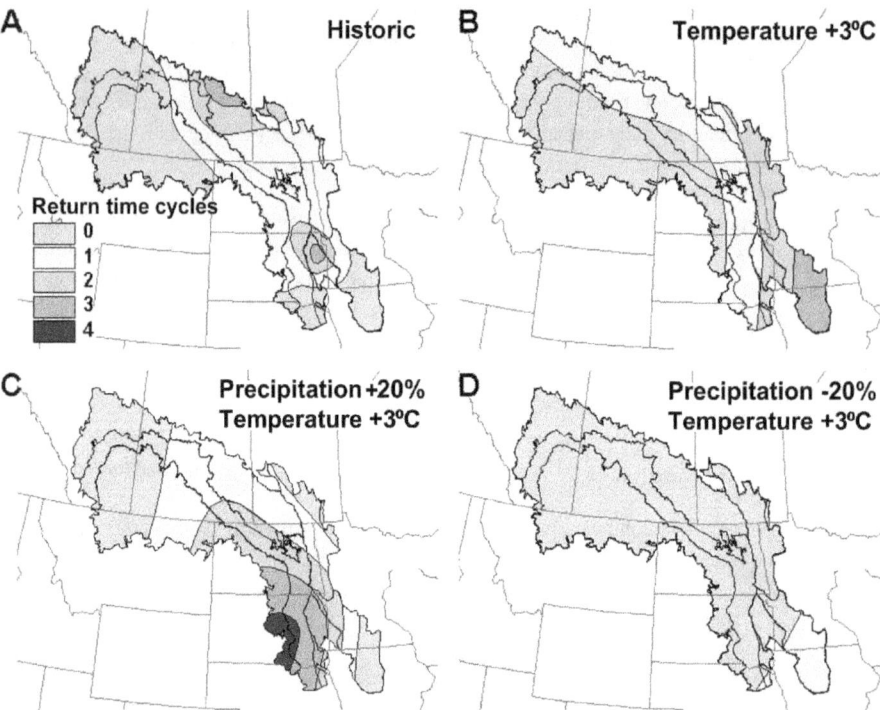

Figure 3.3. Geographic patterns of the speed of the wetland cover cycle, simulated for the Prairie Pothole Region under historic (a) and alternative future (b, c, and d) climatic conditions. (Johnson et al., 2005)

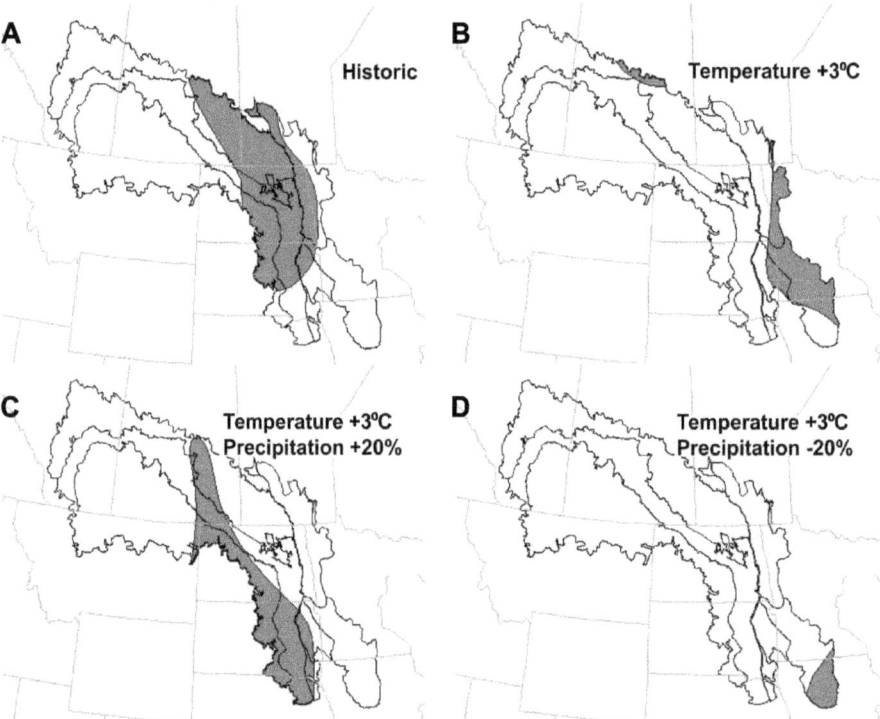

Figure 3.4. Simulated occurrence of highly favorable water and cover conditions for waterfowl breeding (occurrence of at least one return time and hemi-marsh conditions at more than 30 percent frequency) across the Prairie Pothole Region under historic (a) and alternative (b, c, and d) future climatic conditions. (Johnson et al., 2005)

3.3. CASE STUDY 3. BROAD-SCALE FOREST DIEBACK AS A THRESHOLD RESPONSE TO CLIMATE CHANGE IN THE SOUTHWESTERN UNITED STATES

The ecological dynamics of semiarid forests and woodlands in the southwestern United States are observed to respond strongly to climate-driven variation in water availability, with major pulses of woody plant establishment and mortality commonly corresponding to wet and dry periods (Swetnam and Betancourt, 1998). Although human management of these forests is also a factor in tree mortality, it is clear that climate-induced water stress can trigger rapid, extensive, and dramatic forest dieback (Breshears *et al.*, 2005), exemplifying significant ecosystem threshold responses to climate. Broad-scale tree mortality can shift ecotones between vegetation types (Allen and Breshears, 1998) and alter regional distributions of overstory and understory vegetation (Gitlin *et al.*, 2006; Rich *et al.*, 2008). Rapid forest dieback also has nonlinear feedbacks at multiple spatial scales with other ecological disturbance processes, such as fire and erosion (Allen, 2007), which, in some cases, leads to additional nonlinear threshold behaviors. Massive forest mortality is an example of a threshold phenomenon with substantial implications for future ecosystem dynamics and management of lands undergoing such changes (Millar *et al.*, 2007).

Assessments of potential global change impacts initially focused on how vegetation types matched given climatic envelopes (IPCC, 1996). Subsequent research has considered how vegetation patterns might migrate in response to a changing climate with a focus on rates of plant establishment, has documented that forest turnover rates follow global and regional patterns of productivity (significantly driven by climate) (Stephenson and van Mantgem, 2005), and has increasingly moved toward dynamic global vegetation models that try to incorporate more realistic disturbance dynamics (Scholze *et al.*, 2006; Purves and Pacala, 2008). Currently, climate-induced dieback of woody plants is being recognized as an important vegetation response to climate variation and change, with examples of forest dieback emerging from around the world (Allen and Breshears, 2007). [It should also be noted that other recent studies have documented increased tree growth in dry forests, perhaps because of increased water use efficiency (Soule and Knapp, 1999).] Recent research shows that water stress appears to be driving increases in background tree mortality rates in western North American forests (van Mantgem and Stephenson, 2007). In addition, observations of extensive tree dieoff—especially from semiarid ecosystems where woody plants are near their physiological limits of water stress tolerance—are being documented globally, for example, in Australia (Fensham and Holman, 1999), Africa (Gonzalez, 2001), west Asia (Fisher, 1997), Europe (Dobbertin *et al.*, 2007), South America (Suarez *et al.*, 2004), and North America (Breshears *et al.*, 2005). Climate-induced water stress over extended time periods can exceed the physiological tolerance thresholds of individual plants and directly cause mortality through either (1) cavitation of water columns in the xylem conduits ("hydraulic failure") or (2) forcing plants to shut down photosynthesis to conserve water, leading to "carbon starvation" (McDowell *et al.*, 2008). These individual-scale threshold responses to climate stress can trigger tree mortality that propagates to landscape and even regional spatial scales (Allen, 2007), sometimes amplified by biotic agents (like bark beetles) that can successfully attack and reproduce in weakened tree populations and generate massive insect population outbreaks with positive feedbacks that greatly increase broad-scale forest mortality (Kurz *et al.*, 2008).

Ecotones are areas where vegetation changes in response to climate are expected to be most rapid and prominent (Beckage *et al.*, 2008), as highlighted by a southwestern case study of drought effects on vegetation during the 1950s (fig. 3.5; Allen and Breshears, 1998). Severe drought across the southwestern United States during the 1950s caused ponderosa pine (*Pinus ponderosa*) trees at lower, drier sites to die, resulting in an upslope shift of the ponderosa pine forest and piñon-juniper woodland ecotone of as much as 2 kilometers (km) in less than 5 years, producing a rapid and persistent change in dominant vegetation cover. Similarly, within the distributional range for the piñon pine (*Pinus edulis*), many trees at lower or drier sites also died (Swetnam and Betancourt, 1998).

Although tree mortality almost certainly occurred across much of the southwestern United

These individual-scale threshold responses to climate stress can trigger tree mortality that propagates to landscape and even regional spatial scales.

States in response to the 1950s drought (and probably for previous regional-scale droughts as well), few studies exist that allow scientists to test projections about the rapidity and extent of potential vegetation dieoff responses to drought. A recent drought beginning in the late 1990s and peaking in the early 2000s affected most of the western United States. This was the most severe drought in the Southwest since the 1950s. Substantial mortality of multiple tree species has been observed throughout the Southwest during this 2000s drought (fig. 3.6; Gitlin *et al.*, 2006; U.S. Forest Service, 2006; Allen, 2007). For example, mortality of the piñon pine spanned major portions of the species' range, with substantial dieoff occurring across at least 1,000,000 ha from 2002 to 2004 (Breshears *et al.*, 2005; U.S. Forest Service, 2006). For both droughts, much of the forest mortality was associated with bark beetle infestations, but the underlying cause of dieback appears to be water stress associated with the drought conditions.

The precipitation deficit that triggered the recent regional-scale dieoff of the piñon pine across the Southwest was not as severe (dry) as the previous regional drought of the 1950s, but the recent 2000s drought was hotter than the 1950s drought by several metrics, including mean, maximum, minimum, and summer (June-July) mean temperature (Breshears *et al.*, 2005). Although historic data from the 1950s is limited, available data suggest that piñon pine mortality in response to the recent drought has been more extensive, affected greater proportions of more age classes, and occurred at higher elevation and wetter sites than in the 1950s drought. Hence, the warmer temperatures associated with the 2000s drought may have driven greater plant water stress through increased evapotranspirational demand and resulted in more extensive tree dieoff. Because global change is projected to result in droughts under warmer conditions (referred to as "global-change type drought") the severe piñon pine dieback from the recent drought may be a har-

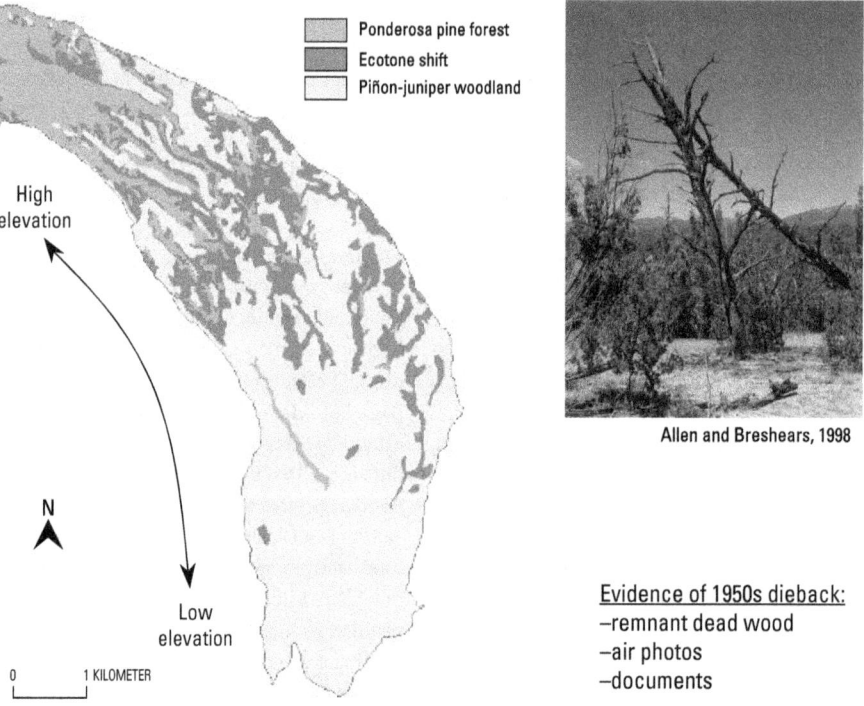

Figure 3.5. Changes in vegetation cover between 1954 and 1963 at Frijolito Mesa, Jemez Mountains, New Mexico, showing the persistent ponderosa pine forest (365 ha), the persistent piñon-juniper woodland (1527 ha), and the ecotone shift zone (486 ha) where forest changed to woodland (from Allen and Breshears, 1998).

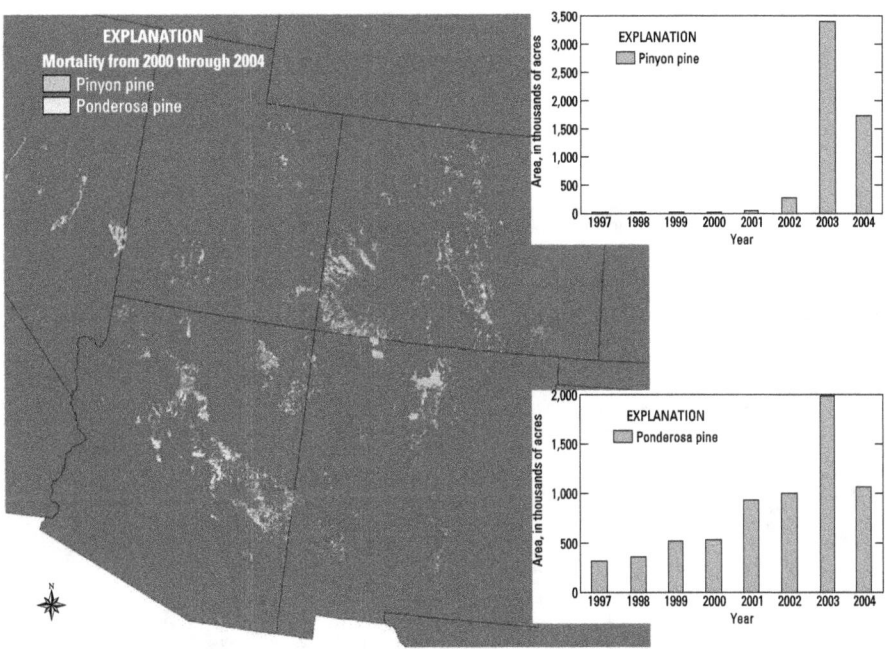

Figure 3.6. Map showing cumulative area dieback for piñon pine (Pinus edulis) and ponderosa pine (Pinus ponderosa) from 200 through 2004 and graphs showing the acreage of dieback from 1997 through 2004 in the Four Corners States of Arizona, New Mexico, Colorado, and Utah. Based upon annual aerial forest insect and disease activity inventories by the U.S. Forest Service.

binger of vegetation response to future global-change type droughts (Breshears *et al.*, 2005).

In addition to the dieoff of dominant overstory tree species, high levels of dieback also were observed in other Southwestern U.S. species and life forms in response to the warm regional drought in the 2000s (Gitlin *et al.*, 2006; Allen, 2007). These include species where bark beetles are unimportant or nonexistent, including one-seed juniper (*Juniperus monosperma*)—a co-dominant with piñon pine for much of its range; shrubs such as wavy-leaf oak (*Quercus undulate*) and mountain mahogany (*Cerco-carpus montanus*); and blue grama (*Bouteloua gracilis*), the dominant herbaceous species in many of these woodland systems.

In addition to direct climate-induced mortality, severe protracted drought also can cause substantial reductions in the productivity and soil surface cover of herbaceous plants, which in turn affects numerous other ecological processes. In particular, reductions in herbaceous ground cover can trigger a nonlinear increase in soil erosion once a threshold of decreased herbaceous cover has been crossed, through increased connectivity of bare soil patches (fig.

3.7; Davenport *et al.*, 1998; Wilcox *et al.*, 2003; Ludwig *et al.*, 2005). On the other hand, dieback of woody canopies tends to cause an immediate successional shift toward greater cover of understory vegetation if moisture conditions are adequate (for example, Rich *et al.*, 2008), which propagates a different set of effects.

Overall, the dieback of overstory vegetation affects numerous key ecosystem processes, which are tied to site-specific distributions of incoming energy and water (Zou *et al.*, 2007), and has multiple cascading ecological effects. Widespread tree mortality may propagate additional pervasive changes in various ecosystem patterns and processes. Breshears (2007) summarizes the important ecological role of woody plant mosaics in semiarid ecosystems:

A large portion of the terrestrial biosphere can be viewed as lying within a continuum of increasing coverage by woody plants (shrubs and trees), ranging from grasslands with no woody plants to forests with nearly complete closure and coverage by woody plants (Breshears and Barnes, 1999; Breshears, 2006).

Climate change has the potential to drive multiple nonlinear or threshold-like processes that can interact in complex ways, making ecological predictions difficult.

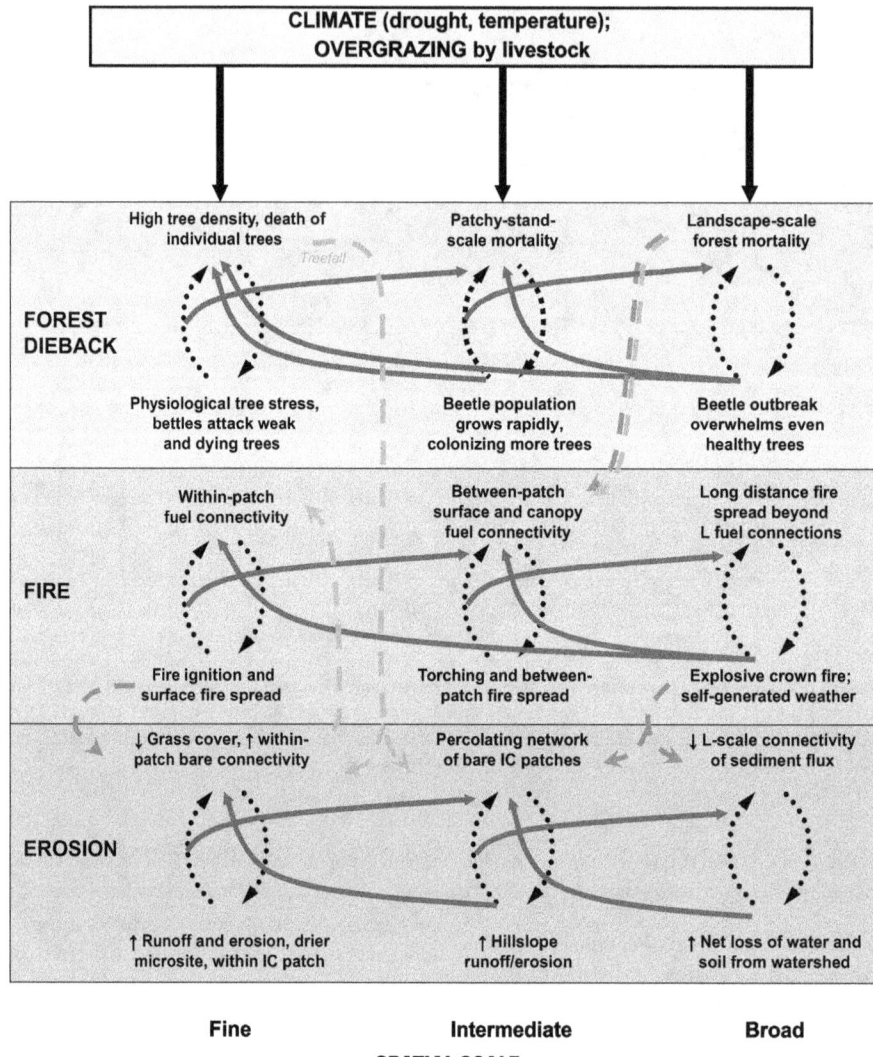

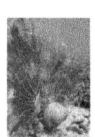

Figure 3.7. Diagram representing interactions across spatial scales for three different disturbance processes (forest dieback, fire, and erosion) in northern New Mexico landscapes (from Allen, 2007). Dashed black arrows represent pattern-process feedbacks within three different spatial-scale domains, with one example of pattern and process shown for each domain for each disturbance. Solid black arrows indicate the overarching direct effects of widespread environmental drivers or disturbances (such as climate and overgrazing) on patterns and processes at all scales. Blue arrows indicate the point at which altered feedbacks at finer spatial scales induce changes in feedbacks at broader scales (for example, fine-scale changes cascade to broader scales), and also where changes at broader scales overwhelm pattern-process relationships at finer scales. Red dashed arrows illustrate some examples of amplifying (positive feedback) interactions between disturbance processes within and between spatial scales; green dashed arrows illustrate dampening (negative feedback) interactions between disturbance processes. Abbreviations: L = landscape; IC = intercanopy (interspaces between tree canopies).

The characteristics of woody plants determine fundamental descriptors of vegetation types including grassland, shrubland, savanna, woodland, and forest. Because woody plants fundamentally affect many key aspects of energy, water and biogeochemical patterns and processes, changes in woody plant cover are of particular concern.

Climate-driven, rapid forest dieback has feedbacks with other ecological disturbance processes, such as fire and erosion, in some cases leading to further nonlinear ecosystem threshold behaviors (fig. 3.7). Warming and drying climate conditions are driving higher-severity fire activity at broader scales in the southwestern United States directly (Swetnam and Betancourt, 1998; Westerling *et al.*, 2006), and probably also indirectly where forest dieback changes fuel conditions (fig. 3.7: Bigler *et al.*, 2005). High-severity stand-replacing fires within woodlands and forests can almost instantly cause large reductions in tree canopies and soil surface covers, thereby also triggering dramatically increased rates of runoff and soil erosion for several years post-fire until vegetation regrowth restores adequate land surface cover (Veenhuis, 2002). Forest dieback, fire, and erosion also have significant effects on ecosystem carbon pools (Breshears and Allen, 2002; Kurz *et al.*, 2008). The combined interactive effects of climate-driven ecological disturbance processes (vegetation dieback, fire, and erosion) are highlighted by the major changes in woodland and forest ecosystems that have occurred in northern New Mexico during the past 60 years (fig. 3.8; Allen, 2007). Climate-induced forest dieback, fire, and accelerated erosion already may be causing permanent type-conversion changes to some southwestern ecosystems. Even without factoring in ongoing or predicted climate changes, it will be at least several decades to centuries before reestablish-

ment of predisturbance tree canopy covers will occur on many semiarid woodland and forest sites in this region (Allen and Breshears, 1998; Savage and Mast, 2005).

Examples of drought-induced tree dieoff in semiarid forests and woodlands highlight the rapidity and extensiveness with which climate stress can trigger pervasive and persistent ecosystem changes. Climate change has the potential to drive multiple nonlinear or threshold-like processes that can interact in complex ways, including tree mortality, altered fire regimes, energy and water budget changes, and soil erosion thresholds (Allen, 2007), making ecological predictions difficult (McKenzie and Allen, 2007). For example, the projections of state-of-the-art dynamic global vegetation models "are currently highly uncertain, making vegetation dynamics one of the largest sources of uncertainty in Earth system models" (Purves and Pacala, 2008). Additional research, including research on threshold responses, is needed to improve projections of the nonlinear ecological effects of expected climate changes, such as broad-scale forest dieback, associated ecosystem dynamics, and effects on carbon budgets and other ecosystem goods and services (Breshears and Allen, 2002; Millennium Ecosystem Assessment, 2005; Millar *et al.*, 2007).

Figure 3.8. Increased herbaceous cover has developed since recent piñon pine forest dieback in the Jemez Mountains of New Mexico and may promote surface fire regimes and changes in runoff and erosion patterns. July 2004.

3.4.   CASE STUDY 4. THRESHOLDS IN CLIMATE CHANGE FOR CORAL-REEF ECOSYSTEM FUNCTIONING

Corals are perpetually subjected to environmental changes in time and space. As adult colonies, corals are sessile, remaining in one location over time, and therefore, are subjected to changes in environmental factors through a temporal scale. As larvae, corals are motile, and each must select a location from a complex and variable array of available sites. Corals are resilient to changes, both spatially and temporally, through acclimatization, adaptation, local environmental ameliorations, initial community composition, and the morphological characteristics of the reef. It is reasonable to assume that most corals will not go extinct with global climate change because of their abilities to acclimatize, to adapt, and to broadcast their larvae over a large-scale landscape (Paulay, 1997). Systems consist of mixtures of positive and negative feedbacks, with positive feedbacks tending to alter the nature of the system, and negative feedbacks tending to minimize these changes (Chapin *et al.*, 1996). The threshold, or tipping point, for coral-reef ecosystems is the point along the environmental gradient at which the ecological or biological processes change from negative feedback for net accretion to positive feedback or reef erosion. When net accretion decreases to a point of net erosion of the reef, the resiliency of the system to return to a functioning coral ecosystem has been greatly reduced, potentially affecting the rate of reaching a threshold of coral mortality. Natural stressors, which are the results of anthropogenic stressors (for example, overfishing, pollutants, sedimentation, habitat destruction), that can lead to positive feedbacks, potentially decreasing the threshold level of coral mortality, include the following (Birkeland, 2004):

- inverse density dependence (or Allee effect);
- algal abundance at levels beyond the capacity of herbivores to keep in balance;
- predation of corals at a rate higher than the rate of recovery and coral population replenishment;
- bioerosion of corals;
- the prevalence of crustose coralline algae, which weakens binding of the substratum, is decreased and thereby decreases successful coral recruitment; and

It is likely that the crossing of thresholds in coral ecosystems began nearly three decades ago with no evidence that the rate of degradation is decreasing.

- invasive species—the establishment of introduced species, which modify the habitat in ways that favor the survival and dominance of the introduced species and displacement of natural species.

Such processes as these stressors and the feedback mechanisms of corals to these stressors have determined the substantial degradation of coral reefs over the past three decades in the tropical western Atlantic Ocean (Gardner *et al.*, 2003) and in the Indo-Pacific Ocean (Bruno and Selig, 2007). It is *likely* that the crossing of thresholds in coral ecosystems began nearly three decades ago with no evidence that the rate of degradation is decreasing (Birkeland, 2004).

Although anthropogenic modification of local ecological processes has been the dominant force in coral-reef degradation (Birkeland, 2004) and tipping points have been crossed decades ago in many areas (Gardner *et al.*, 2003; Bruno and Selig, 2007), global changes in climate and oceanic characteristics are now becoming more apparent. Global processes that are affecting coral reefs are sea-level rise, the decline in pH of seawater, and the increase in seawater temperature, which are related to the increased concentration of atmospheric CO_2.

Rise and Fall of Sea Level.—Coral reef ecosystems have experienced rise and fall of sea levels several times in geological history with associated effects on reef functioning (Hallock, 1997) (with "reef functioning defined as constructing reefs upwardly). Reef accretion has stopped for periods of time in excess of 10 million years (Copper, 1994; Hallock, 1997), the threshold for the cessation of reef upward growth being the time of decreasing sea level (Hallock, 1997; Hubbard, 1997). It is hard to determine the effect of climate change alone on whether corals will keep pace with sea level rise, increasing water temperatures, and change in ocean pH. The rate of sea level rise alone does not provide a predictable tipping point for reef deposition that can be generalized over a region (Hallock *et al.*, 1993; Kleypas *et al.*, 2001; Garrison *et al.*, 2003). Whether coral reefs keep up with sea level rise depends on a multitude of local environmental factors and the degree to which these factors stress the corals themselves, which will affect the rate at which the threshold for coral mortality will be reached (Hubbard, 1997).

Decrease in Seawater pH.—The concentration of CO_2 in the atmosphere is generally expected to reach two times the preindustrial (late 18th century) levels by 2065 (Houghton *et al.*, 1996). As CO_2 concentration increases in the atmosphere, the surface seawaters take up more CO_2. The increased uptake of atmospheric CO_2 by the surface waters of the ocean leads to a decrease in pH of surface waters, an increase in the proportion of bicarbonate ions (HCO_3^-), and a decrease in the proportion of carbonate ions (CO_3^{2-}) (Feely *et al.*, 2008). The overall effect on the rate of precipitation of coral skeleton is negative.

$$CO_2 + H_2O \Leftrightarrow HCO_3^- + H^+ \Leftrightarrow CO_3^{2-} + 2H^+$$

The oceans have already taken up an additional one-third to one-half of industrial-age emissions of CO_2 (fig. 3.9), and the concentrations of carbonate ions in the oceans have decreased from 11 percent (preindustrial), to 9 percent (now) and are projected to decrease to 7 percent when carbonate concentrations are double the preindustrial concentrations, perhaps in 3 to 5 decades (ISRS, 2007).

Kleypas *et al.* (1999) determined that doubled atmospheric CO_2 will lead to a 14 percent to 30 percent decrease in reef calcification rates. This was estimated to be a general threshold from net carbonate accretion to net carbonate loss by Kleypas *et al.* (2001). Net reef accretion

is potentially reduced to zero when increased CO_2 in the atmosphere reaches about 500 to 600 ppm. On the other hand, CO_2 is less soluble in seawater at higher temperatures. While increased concentrations of atmospheric CO_2 may be accelerating the uptake of CO_2 by surface seawater, global warming may be slightly damping the uptake. But of more substantial influence in accelerating the tipping point of net reef accretion are the synergistic biological effects on corals of reduced growth in the face of natural and anthropogenic stressors.

Sabine *et al.* (2004) showed that uptake of anthropogenic CO_2 by subtropical Atlantic waters has been greater than by Pacific waters. The north Atlantic occupies only 15 percent of the world's total ocean area and stores 23 percent of the total anthropogenic (fossil-fuel and cement-manufacturing emissions) CO_2 taken up by the world oceans. Pacific waters are less receptive to the uptake of CO_2 and therefore are buffered from a decrease in pH because of higher concentrations of dissolved inorganic carbon. As seawater becomes warmer, coral reef net accretion will probably become slightly more restricted in latitude (Kleypas *et al.*, 1999, 2001) because of the changes in chemistry from CO_2 uptake in the world's oceans.

Studies have shown that the resilience of corals to lower pH of ocean waters decreases with input of nutrients from continents. Anne Cohen of the Woods Hole Oceanographic Institution has taken core samples from 20 large *Diploria*

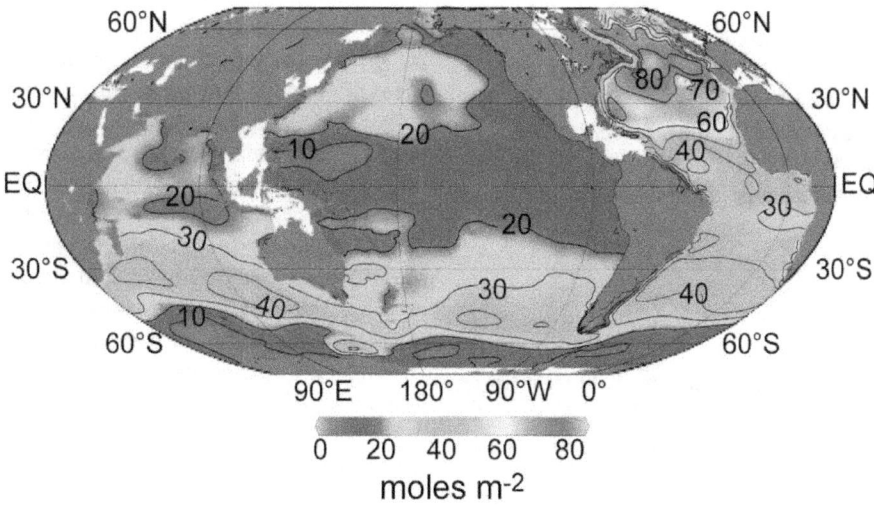

Figure 3.9. Status of oceanic uptake of CO_2. Source: Sabine *et al.*, 2004.

labyrinthiformis colonies in Bermuda and found that the rate of calcification has significantly declined since 1959 (Cohen *et al.*, 2008). This is consistent with the decrease in pH of the ocean waters of the northern Atlantic (Sabine *et al.*, 2004) and the concomitant lowering of the saturation level of aragonite in coral skeletons. The corals are, nevertheless, doing well, and the coral-reef ecosystem is intact in Bermuda (Murdoch *et al.*, 2008), which is relatively distant from Continental land masses. In contrast, the coral reefs have been degrading for decades in the Caribbean and western Atlantic (Gardner *et al.*, 2003), which are close to continental land masses and associated land-surface runoff. (The input of fixed nitrogen from excess fertilizer runoff from the Mississippi River into the western Atlantic has averaged 1.6 million metric tons per year since the 1980s and the input of phosphate has averaged a hundred thousand metric tons per year).

Done *et al.* (in press) report that coral communities on the Great Barrier Reef have been losing their resilience since about 1997. Done *et al.* found that loss in resilience on the Great Barrier Reef is correlated with nutrient (fixed nitrogen) input.

A number of studies presented at the 11th International Coral Reef Symposium reported that coral-reef systems are still resilient in areas far from continental land masses (for example, the Andaman and Maldive Archipelagoes, Chagos and Maldives in the Indian Ocean, Moorea, Fiji and American Samoa in the Pacific, and Bermuda in the Atlantic).

Seawater Warming.—The thresholds in tolerance of corals to an increase in water temperature and its duration before "bleaching" (expelling the symbiotic *zooxanthellae*) is predicted by the degree heating week (DHW) record (a NOAA satellite-derived experimental product), 12-week accumulations measured as °C weeks. The DHW product is an accumulation of hotspot values over the bleaching threshold (1°C above the maximum monthly mean). The threshold values of DHW vary from site to site because the maximum monthly mean varies from site to site; thus, corals are likely adapted to their own threshold temperatures at each site. Furthermore, the past history of events in the physical environment and local characteristics of the physical environment can modify

the actual location of the threshold or tipping point (Smith and Birkeland, 2007). Based on our knowledge of tolerances and the gaps in the literature on thresholds identified in developing this SAP, corals are *likely* to reach a threshold with an increase in sea water temperatures.

Mechanisms of Reef Resilience That Alter Thresholds.—The resilience of corals to environmental changes is largely determined by their capacity to acclimatize (adjust physiologically and behaviorally). The thresholds of resilience of corals to environmental factors, such as water temperature and ultraviolet (UV) radiation, are altered by changes in symbiotic interactions. Reef-building corals are dependent on symbiotic dinoflagellate algae (*zooxanthellae*) in their endodermal cells for their nutrition and proficiency in deposition of skeleton. There are a number of clades or types of *zooxanthellae*, and the physiological and ecological attributes of zooxanthellae vary among clades (Abrego *et al.*, 2008; Berkelmans and van Oppen, 2006; McClanahan *et al.*, 2005; Baker *et al.*, 2004; Baker, 2004; Buddemeier *et al.*, 2004; Rowan, 2004; Baker, 2003; Rowan and Knowlton, 1995). The symbiotic relationship breaks down under stressful conditions of extra warm seawater or strong UV radiation. Under these conditions, corals sometimes expel much of the zooxanthellae of clade C and allow the buildup of clade D, with which the coral growth rate is slower but survival under stressful conditions may be greater. As with morphological adjustments, the symbiotic adjustments of corals may be determined by a balance between the stresses imposed by the physical environment and by ecological interactions with other species (Bruno and Selig, 2007). In addition to adjustments in morphology and symbiotic relationships, acclimatization can occur through biochemical conditioning where increased water temperature triggers a substantial increase in biochemical activity in corals. Intense biochemical activities (such as the increase in the amounts of heat shock proteins and ubiquitin produced) resulting from changes in water temperature, may indicate processes of biochemical conditioning and acclimatization that might increase the resilience of the coral from increased seawater temperature (that is, increase the threshold level of coral mortality) (Smith and Birkeland, 2007).

Whether changes in morphology, symbiotic relationships, physiological conditioning, or

Whether changes in morphology, symbiotic relationships, physiological conditioning, or production of biochemicals are the mechanisms to shift the threshold for survival from climate change, acclimatization costs the coral in terms of energy and materials that would otherwise be available for growth and successful competition.

Figure 3.10. Branching corals overgrowing mound-shaped corals.

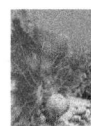

production of biochemicals are the mechanisms to shift the threshold for survival from climate change, acclimatization costs the coral in terms of energy and materials that would otherwise be available for growth and successful competition. Acclimatization in corals can occur either as an accumulation of a simultaneous array of biochemical mechanisms to resist stress (robustness) or as an array of alternative paths of development or symbiotic associates (plasticity). The mound-shaped species of *Porites* (such as *P. lobata*) are robust and live in a wide range of habitats. They are the last to drop out of the coral community near a river mouth or in bays with increasing turbidity. Species of *Acropora* dominated the reef front at the municipal sewer outfall for Koror, Palau, until predation on corals by the crown-of-thorns starfish and bleaching by the large-scale seawater warming of 1997–98 killed the *Acropora* spp. but not the *Porites* spp. (Richmond *et al.*, 2002). *Porites* can maintain itself rather constantly despite fluctuations in the external physical environment, but at a metabolic cost (fig. 3.10).

The relatively rapidly growing *Pocillopora eydouxi* display plasticity and can differ substantially among habitats in rates of growth, colony morphology, and types of zooxanthellae hosted. *Pocillopora* are generally more vulnerable to the physical environment and so their growth rates vary among habitats and they are more likely to bleach (expel zooxanthellae and/or photosynthetic pigments) with higher than usual water temperatures and with more intense UV radiation.

Factors That Shift the Thresholds.—Corals are most vulnerable to infrequent or very frequent environmental changes. Corals can acclimatize (physiological or behavioral response) or adapt (genetic response) to environmental changes of intermediate frequency. (Adaptation is genetic change in a population in response to natural selection). If the phenomena, such as extraordinarily warm seawater, are infrequent enough to be unpredictable, corals will not be able to acclimatize or adapt, and if too frequent, will not have time to recover between events, thus decreasing the threshold level of coral mortality (Smith and Birkeland, 2007).

The factor of duration relates to the different effects of acute and chronic disturbances on the resilience of coral communities. The threshold seawater temperature associated with global climate change is determined in part by the duration of the warm water event. In 1997–98, an increased average surface seawater temperature of 1.0° to 1.5°C (to about 30° or 31°C) over a period of several weeks caused extensive mortality of corals in the Indian Ocean, the southwestern Pacific Ocean, and the western Atlantic Ocean (Bruno and Selig, 2007). In contrast, daily fluctuations of 6°C to 6.5°C (to about 34°C or 35.5°C) in reef flat pools in American Samoa are endured in good health by about 80 species of corals.

The threshold seawater temperature that severely affects a coral will be higher in areas of constant or even intermittent strong water motion and the threshold of temperature tolerance will be lower in areas of weak water motion (Smith and Birkeland, 2007). Thresholds in levels of tolerable input of nutrients or sediment will be low in backwaters and relatively much higher in areas of strong current (Smith and Birkeland, 2007; Garrison *et al.*, 2003). In contrast, it will take substantially longer for the ecosystem to solidify rubble into a stable substratum for reef recovery in areas of strong water motion than in areas of low water motion. The threshold of tolerance of corals to infection by disease is sometimes lowered by stress from other environmental factors and by abrasion of surface tissue by predators or other objects (Garrison *et al.*, 2003). The physical and biological environments are a complex system of factors that potentially act synergistically to shift the threshold of the specific factor associated with climate change.

Thresholds.—Thresholds should be considered at two stages: the first at which the population is killed or the ecosystem becomes dysfunctional, and the second at which the population or the ecosystem is prevented from becoming reestablished. An acute disturbance to a coral reef is a distinct event whereas a chronic disturbance is an ongoing process. Coral-reef communities

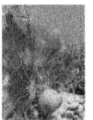

in the Pacific (American Samoa) have been severely affected by large-scale acute disturbances, such as outbreaks of the coral-eating crown-of-thorns starfish *Acanthaster planci* (1938, 1978), hurricanes (1981, 1987, 1990, 1991, 2004, 2005), and bleaching in response to seawater warming (1994, 2002, 2003). This is in contrast to the western Atlantic where there has been chronic disturbance resulting in degradation of coral reef systems for a half a century (Gardner *et al.*, 2003). When allowed a 15-year interval between acute disturbances, the Pacific coral communities have recovered (Birkeland *et al.*, 2008). Whereas in the relatively small area of the tropical western Atlantic, external stressors such as nutrients (Hallock *et al.*, 1993), pollutants (Garrison *et al.*, 2003), and diseases (Lessios *et al.*, 1984) from wide-scale events on continents (Hallock *et al.*, 1993; Garrison *et al.*, 2003) can disperse across the entire region. This chronic disturbance decreases the threshold of coral mortality. A recent paper by Bruno and Selig (2007) reported that 3,168 square kilometers of reef has been dying each year rather uniformly throughout the Indo-Pacific Ocean. Reefs are appearing to be losing their resilience globally.

Coral reefs in the Pacific (America Samoa) have managed to maintain resilience because disturbances have been acute events. External stressors from overfishing, however, have been chronic, and the fish communities have not been as resilient as the corals (Zeller *et al.*, 2006a, b). Thresholds of coral reef systems need to take into account the whole system and not just the corals to ensure a resilient and adaptive system in the face of climate change.

Examples of Threshold Change in Ecosystems

Lead Authors: Daniel B. Fagre, USGS; Colleen W. Charles, USGS.
Authors: Craig D. Allen, USGS; Charles Birkeland, USGS-University of Hawaii; F. Stuart Chapin, III, University of Alaska; Peter M. Groffman, Institute of Ecosystem Studies; Glenn R. Guntenspergen, USGS; Alan K. Knapp, Colorado State University; A. David McGuire, USGS-University of Alaska; Patrick J. Mulholland, Oak Ridge National Laboratory; Debra P.C. Peters, USDA Agricultural Research Service; Daniel D. Roby, USGS-Oregon State University; George Sugihara, Scripps Institute of Oceanography and University of California at San Diego.
Contributing Authors: Brandon Bestelmeyer, Jornada Basin LTER; Julio L. Betancourt, USGS; Jeffrey E. Herrick, Jornada Basin LTER

4.1. BACKGROUND

The existence of ecological thresholds has long been apparent to people who depend on natural resources. Fisheries collapses, for instance, have been noted for centuries. However, ongoing climate change has given this issue greater urgency because more ecosystems may be getting pushed toward response thresholds simultaneously, and based on gaps in the literature identified through the development process for this assessment (SAP 4.2), little is known regarding where the tipping points are.

Ecosystems are very likely to differ significantly in their potential for climate change to impact them to the point that thresholds are crossed and substantial alterations occur. Given the magnitude and pervasiveness of climate change, it is surprising how little is known regarding the sensitivity of different ecosystems to any single aspect of climate change (such as increased temperatures), and even less is known about the impacts of multiple climate change factors. This lack of basic understanding represents a critical knowledge gap and research challenge, one that is further complicated by the fact that climate change is only one component of global change and that multiple alterations to climate, biogeochemical cycles, and biodiversity are occurring in tandem.

Summarized below are examples of where ecological thresholds have been crossed; they are less detailed than the case studies of chapter 3 but represent different geographic areas, ecosystem types, and drivers of change. These examples include the new stressor of climate change and reflect how it leads to new ecosystem responses. For example, the temperature increases documented for many areas can likely cause an ecosystem changeover when normal droughts are experienced because the additional evapotranspiration demand of higher temperatures exceeds the tolerance capacity of trees, leading to the massive forest dieback described in case study 3.

4.2. EXAMPLE OF THRESHOLDS FROM THE PAST

Thresholds appear to have been crossed in the past, leading to ecosystem changes that persist today. A recent example of threshold behavior is the encroachment of woody plants into perennial grasslands that has occurred throughout arid and semiarid regions of the world for at least the past several centuries. This broad-scale land-cover conversion and associated soil degradation (that is, desertification) has local to global consequences for ecosystem services, such as reduced air and water quality (Schlesinger *et al.*, 1990; Reynolds and Stafford Smith, 2002). Multiple interacting processes and threshold behavior are involved in these dynamics (Rietkerk and van de Koppel, 1997).

Given the magnitude and pervasiveness of climate change, it is surprising how little is known regarding the sensitivity of different ecosystems to any single aspect of climate change, and even less is known about the impacts of multiple climate change factors.

Cross-scale linkages among local soil and grass degradation, landscape connectivity of erosion processes, and land-cover and weather feedbacks have been invoked to explain threshold behavior in space and time that occur during desertification (Peters *et al.*, 2006). Four stages and three thresholds have been identified as the spatial extent of desertified land increases through time (Peters *et al.*, 2004). Following introduction of woody plant seeds into a grass-dominated system (Stage 1), local spread often occurs as a result of feedback mechanisms between plants and soil properties interacting with wind and water erosion to produce fertile plant islands surrounded by bare areas that move the system across a threshold into Stage 2 (Schlesinger *et al.*, 1990). This rate of spread may be slower than other stages as a result of interactions between plant life history characteristics that occur infrequently, such as recruitment, and the low precipitation and high temperatures that characterize dry regions. As the size and density of woody plants increase through time, contagious processes among patches, primarily wind and water erosion that connect bare soil patches, become the dominant factors governing the rate of desertification. As a result, a nonlinear increase in woody plant cover occurs and a second threshold is crossed as the system enters Stage 3. Through time, sufficient land area can be converted from grassland (low bare area, low albedo) to woodland (high bare area, high albedo) so that regional atmospheric conditions, in particular wind speed, temperature, and precipitation, are affected. At this point, a third threshold is crossed where land-atmosphere interactions with feedbacks to vegetation control system dynamics (Stage 4) (Pielke *et al.*, 1997). Feedbacks to broad-scale vegetation patterns have been documented in the Sahara region of Africa (Claussen *et al.*, 1999).

4.3. TEMPERATURE INCREASES ARE PUSHING ECOSYSTEMS TOWARDS THRESHOLDS

The impacts of increasing temperatures resulting from climate change are not independent of the effects of other important environmental stressors, and thus, need to be assessed in the context of multiple, interacting stressors. The IPCC (2007) reports with very high confidence that the increased warming effect of climate change is strongly affecting natural biological systems in both marine and freshwater systems. The chemical and physical characteristics of lakes experience major effects owing to changes in temperature, especially changes in nutrient dynamics. Increased temperatures in lake systems will affect the distributions, growth, and survival of fish and many other aquatic organisms. Tied with increased temperatures is a change in precipitation, which can cause substantial physical and chemical changes in lakes, streams, and wetlands (as discussed in chapter 3) with large consequences for aquatic biota. In marine systems, increased temperature from climate change is affecting coastal resources and habitats because of sea-level rise that is caused by thermal expansion of the oceans and the melting of ice cover. It also is affecting the strongly coupled atmospheric and oceanic circulation that underpins ecosystem dynamics in wind-driven upwelling shelves and ecosystem susceptibility to modulations of upwelling wind stress causing present day global distribution of shelf anoxia (Chan *et al.*, 2008). This has the potential to affect the rate at which the threshold for mortality will be reached for demersal fish and benthic invertebrate communities in these shallow waters. The rate of sea-level rise is expected to accelerate because of global warming. Salt marshes, which must increase their vertical elevation at rates that keep pace with sea-level rise or risk transformation to a lower position along the marsh gradient, may experience a change of marsh type. Transition from one type of marsh to another (for example, high marsh to low marsh) at a given point has been described as ecosystem state change (Miller *et al.*, 2001).

The effects of temperature increases on terrestrial systems are further emphasized in the IPCC Assessment Report for Working Group II (IPCC, 2007), where it is stated with very high confidence that the overwhelming majority of studies of regional climate effects on terrestrial species reveal consistent responses to warming trends, including poleward and elevational range shifts of flora and fauna. Responses of terrestrial species to warming across the Northern Hemisphere are well documented by changes in the timing of growth stages (that is, phenological changes),

The overwhelming majority of studies of regional climate effects on terrestrial species reveal consistent responses to warming trends, including poleward and elevational range shifts of flora and fauna.

especially the earlier onset of spring events, migration, and lengthening of the growing season. Changes in abundance of certain species, including limited evidence of a few local disappearances and changes in community composition over the last few decades, have been attributed to climate change. A further indication of effects of increased temperatures is revealed in earlier snowmelt and stream runoff, which affects both aquatic and terrestrial ecosystems and species. Diminished snowpacks that melt earlier in the spring have affected the timing and extent of seasonal wetlands where amphibians breed. A threshold may occur wherein the reduced amphibian population cannot accommodate the necessary shift in the timing of breeding or cannot survive multiple dry years, causing local extinction (Corn, 2003).

There is a need to better understand the complexities of ecosystems and the drivers of change within them and to be able to identify the thresholds of these changes in a changing climate.

4.3.1. Climate Interactions Drive Ecosystems to Thresholds

As important as the increases in temperatures and changes in moisture availability are for causing ecosystems to go through thresholds, it is the interactions that are key to driving the change. In general, plants in undisturbed ecosystems are at their moisture-limited capacity for net primary productivity. Therefore, increased temperatures *and* droughtiness will combine to produce severe stress on plant growth, whereas increased temperatures and increased moisture availability will lessen the stress or may promote plant productivity, leading to an ecosystem with increased resilience. Because evapotranspiration demands on vegetation increase with temperature, thresholds are more likely to occur whenever moisture availability does not simultaneously increase with warming temperatures. The exception is ecosystems that are primarily limited by temperature, such as arctic and alpine ecosystems. In these latter cases, ample moisture means that vegetation can respond without evapotranspiration limits but that threshold changes can still occur as competitive relationships are altered between plant species (Hansell *et al.*, 1998).

4.3.2. Climate Variability Increases Likelihood of Threshold Shifts

The climate drivers that produce threshold ecosystem responses may be complex and involve the interaction of variability in phenology and weather episodes. The "2007 spring freeze" in the Eastern United States is an example. A very warm late winter/early spring period in much of the Southeastern United States in 2007 led to bud-break and development of forest canopy 2 to 3 weeks earlier than usual. A very cold Arctic air mass spread across much of the Eastern United States in early April (an event not unusual for that time of year), dropping the low daily temperatures well below freezing for several days. The freeze killed newly formed leaves, shoots, and developing flowers and fruits and resulted in a sharp drop in vegetation greenness (NDVI) across a large swath of the southeast. The severity of impact was species specific; but at one site affected by this episode, there was a significant reduction in forest photosynthetic activity for at least several weeks after this event, and the leaf-area index was depressed throughout the summer (Gu *et al.*, 2008). Smaller forest leaf area resulted in increased light availability to the stream draining the forest at this site throughout the late spring and summer, leading to increased primary and secondary productivity and higher rates of nutrient uptake and retention in the stream (Mulholland *et al.*, in press). Long-term climate records at this site showed that average late-winter temperatures are increasing but the date of the last hard freeze remains variable and shows no trend with time, suggesting that this heretofore unusual weather event may become more common in the future, which could lead to significant effects on forests and streams. While our understanding of the long-term effects of this episode are unclear, they may *likely* include significant changes in forest composition due to mortality and/or increased susceptibility to pests of the more susceptible species if similar episodes occur in the future (IPCC, 2007).

4.3.3. Other Human Stressors and Climate Change

The interaction of human stresses on ecosystems (for example, land-use change) and climate change may be most evident for lotic ecosystems (those of rivers, streams, and springs) and may produce threshold responses that each stress alone would not produce. Flow

variability over time and space is a fundamental characteristic of lotic ecosystems. It is this temporal and spatial flow variability that defines and regulates biotic composition and key ecosystem processes in streams and rivers (Poff *et al.*, 1997; Palmer *et al.*, 2007). Climate change will alter flow regimes and generate changes to biotic communities in many of these ecosystems, although it is not clear that these flow alterations will produce threshold-type responses in these systems that have evolved in response to high flow variability. However, growing water demands combined with climate-change-induced increases in the severity and duration of droughts in the western United States will likely lead to hydrologic regime shifts in many drainage basins (Barnett *et al.*, 2008).

Recent empirical evidence suggests that severe droughts can produce more dramatic and long lasting effects (for example, loss of biodiversity) on the biological communities of streams and river ecosystems than do other changes in the flow regime, such as floods (Boulton *et al.*, 1992; Lake, 2003). Studies of drought effects on macroinvertebrates in Australian streams where drought is a common and widespread phenomenon suggest that there may be a significant lag effect that prevents recruitment after drought conditions end (Boulton, 2003). Historical evidence exists of large shifts in river fish communities in response to decades-to-century-scale droughts in the Colorado River basin at the end of the Pleistocene (Douglas *et al.*, 2003), but recent findings indicate large uncertainties in long-term effects of drought on fish (Matthews and Marsh-Matthews, 2003).

Many of the expected changes to flow regimes from climate change are similar to those that result from urbanization and other human alterations of drainages. Among these are increased flashiness of hydrographs and longer periods of low or intermittent flow, higher water temperatures, and simplified biotic assemblages (Paul and Meyer, 2001; Roy *et al.*, 2003; Allan, 2004; Nelson and Palmer, 2007). The increases in urbanization that have occurred and are likely to continue in many regions of the United States will very likely exacerbate the effects of climate change.

The strongest evidence for potential threshold effects in rivers and streams appears to be the result of combined impacts of high or increasing human water withdrawals and the likelihood of more frequent or longer droughts under a warming climate. Defining a water stress index equivalent to total human water use divided by river discharge, Vorosmarty *et al.* (2000) showed that the combination of projected population and climate change results in substantial increases in water stress over large areas of the eastern and southwestern United States. In an analysis of sustainable water use in the United States, the Electric Power Research Institute (EPRI, 2003) reported that total freshwater withdrawal exceeded 30 percent of available precipitation over much of the semiarid and arid regions of the United States and over large areas of Florida and other metropolitan areas in the east. High rates of human water use reduce flow and extend low flow periods, restricting and degrading habitat for river and stream biota. Using two scenarios from the 2001 IPCC report, Xenopoulos *et al.* (2005) reported that the combination of climate change and increased water withdrawal may result in loss of up to 75 percent of the local fish biodiversity in global river basins.

There are several examples of potential large-scale threshold responses to the combined effects of human water management and climate-induced drought. In the Columbia River basin of the Pacific northwest, multiple stressors (including population growth; conflicts between hydropower, agriculture, and recreation interests; and ineffective water management institutions and structures) have increased the vulnerability of water resources (Payne *et al.*, 2004; Miles *et al.*, 2007) that were already vulnerable as a result of reduced winter snowpack (Barnett *et al.*, 2005), which generates much of the summer flow, and sustained or repetitive droughts projected by climate change models that would drive water supplies to extreme low levels. Because salmon populations are under considerable stress due to dams, water withdrawals, and other human actions, reduced summer flow under a warmer climate may exceed population sustainability thresholds (Neitzel *et al.*, 1991).

The Colorado River supplies much of the water needs of a large area of the western United States and northern Mexico. The lower portions of the river have become highly vulnerable to drought due to increased demand from population increases. A long-term drought, beginning in about 2000, has lowered water levels considerably in Lakes Powell and Mead, and many climate models project future conditions that will eventually lead to the drying up of Lake Powell and reduced flow in the Colorado River by more than 20 percent. Water allocations for maintaining the ecological integrity of natural communities could drop below thresholds that ensure their viability as scarce water is prioritized for human communities (Pulwarty *et al.*, 2005).

Even in the humid southeastern United States, the combined effects of increased water withdrawals and climate change may exceed thresholds in ecosystem response. The Chattahoochee-Apalachicola River basin in Alabama, Florida, and Georgia is both an important water source for agricultural, industrial, and municipal uses and an important fishery. More than 75 percent of the fish species inhabiting this river system depend on access to floodplain and tributary areas to forage and spawn, and there are flow thresholds below which fish cannot move into these critical areas (Light *et al.*, 1998). Analysis of projected future water withdrawals and climate change for the Chattahoochee-Apalachicola River basin indicates that by 2050, minimum flows will drop below these minimum flow thresholds for at least 3 months in summer in some areas (Gibson *et al.*, 2005). This situation will be exacerbated by the increased percentage of flow that is wastewater effluent combined with lower minimum flows in this rapidly urbanizing basin. This will increase biological oxygen demand and reduce dissolved oxygen concentrations potentially below threshold levels required by some species of fish (Gibson *et al.*, 2005).

The drying up of streams and wetlands represents thresholds that involve contraction or elimination of entire aquatic ecosystems. Prairie rivers, streams, and wetlands of the Great Plains may be particularly vulnerable to these types of thresholds because of the combined effects of water withdrawals for agricultural and municipal uses and projected climate changes that will result in longer periods of drought (Johnson *et al.*, 2005). For example, since the late 1970s, the Arkansas River and many of its tributaries in Kansas have had long periods of dry channels because of extensive surface and groundwater use in its drainage basin (Dodds *et*

al., 2004). The drying up of headwater streams and even some larger streams and rivers for extended periods may become common in wetter areas of the United States as well, particularly as a result of the combined effects of increased water withdrawal and climate change.

Riparian ecosystems are also vulnerable to drought-related thresholds, particularly in the more arid regions of the United States. Riparian forests dominated by cottonwood are being replaced by drought-tolerant shrubs along some rivers in the western United States. Increased surface and groundwater withdrawals combined with drought have resulted in the replacement of riparian forests of native cottonwood (*Populus fremontii*) and willow (*Salix gooddingii*) by an invasive shrub (*Tamarix ramosissima*), resulting in reduced animal species richness, diversity, and abundance over extensive areas along the San Pedro River in Arizona (Lite and Stromberg, 2005). Surface flow and the depth to groundwater appear to be the primary controls on riparian vegetation, with loss of native riparian communities when rivers and streams drop below flow permanence thresholds of 50 to 75 percent (Stromberg *et al.*, 2005, 2007).

4.3.4. Ecosystem Vulnerability and Climate Change

Some ecosystem attributes may be particularly important in generating differential ecosystem vulnerability to climate change, including the likelihood that important thresholds of response are crossed. For example, most ecosystems have a single or just a few dominant species that mediate ecological processes, control the majority of the resources (including space), and/ or have disproportionate impacts on species interactions. Thus, if climate change favors a new dominant species, the prediction is that it will likely be the rate at which the extant species can be replaced and the traits of these new species that will determine the likelihood that the ecosystem will be altered significantly to result in threshold behavior in state or function. For example, ecosystems dominated by long-lived species (for example, trees) with slow population turnover would be expected to be relatively slow to respond to climate change, whereas those ecosystems dominated by short-lived species (for example, annual plants) are expected to be more vulnerable to experiencing substantial change if the new dominant species replacing the old have very different species traits.

The strongest evidence for potential threshold effects in rivers and streams appears to be the result of combined impacts of high or increasing human water withdrawals and the likelihood of more frequent or longer droughts under a warming climate.

Ecosystems can differ dramatically in the sizes of key carbon and nutrient pools, as well as rates of biogeochemical transformations and turnover. These attributes may also determine the rate and magnitude of ecosystem response to climate change if climate forcings influence these biogeochemical attributes. For example, ecosystems with large nutrient pools and/or slow turnover rates are expected to respond minimally to climate-change-induced alterations in nutrients. In contrast, ecosystems with limited nutrient pools and rapid biogeochemical cycling are expected to be more vulnerable to climate change that results in critical thresholds being crossed. The general hydrologic balance of ecosystems would similarly impact ecosystem sensitivity to any climate change that affects water availability. In general, those ecosystems with a ratio of precipitation-to-potential evapotranspiration that is near or below 1:1 will be predicted to be more vulnerable to change than ecosystems where this ratio is greater than 1:1.

Levels of biodiversity (functional traits and species) within an ecosystem may also be important in influencing sensitivity to climate change (Grebmeier *et al.*, 2006). The number and traits of species may buffer ecosystems from change and influence the extent to which immigration of new species will occur. For example, depending on how well species in an ecosystem functionally complement each other and the ability of species to compensate for the change resulting from the loss of the dominant species, the replacement of a dominant species by another species could result in no change or large changes in ecosystem state. Similarly, invading species may result in the rapid crossing of thresholds or may have little or no impact depending on the traits of these species relative to the traits of native species.

Finally, interactions with the natural disturbance regime inherent in an ecosystem, other climate change factors, and other global changes, such as habitat fragmentation and species invasions, will more than likely influence whether or not ecosystems cross response thresholds and experience substantial amounts of change in their structure and function. For example, ecosystems that are historically prone to fire may experience more frequent fires with climate change, making them more susceptible to invasions by exotic species as resources become available postfire.

Interactions with the natural disturbance regime inherent in an ecosystem, other climate change factors, and other global changes will more than likely influence whether or not ecosystems cross response thresholds and experience substantial amounts of change in their structure and function.

CHAPTER 5

What Can Be Done?

Lead Authors: Daniel B. Fagre, USGS; Colleen W. Charles, USGS.
Authors: Craig D. Allen, USGS; Charles Birkeland, USGS-University of Hawaii; F. Stuart Chapin, III, University of Alaska; Peter M. Groffman, Institute of Ecosystem Studies; Glenn R. Guntenspergen, USGS; Alan K. Knapp, Colorado State University; A. David McGuire, USGS-University of Alaska; Patrick J. Mulholland, Oak Ridge National Laboratory; Debra P.C. Peters, USDA Agricultural Research Service; Daniel D. Roby, USGS-Oregon State University; George Sugihara, Scripps Institute of Oceanography and University of California at San Diego.
Contributing Authors: Brandon Bestelmeyer, Jornada Basin LTER; Julio L. Betancourt, USGS; Jeffrey E. Herrick, Jornada Basin LTER

Because there is significant potential for abrupt or threshold-type changes in ecosystems in response to climate change, what changes must be made in existing management models, premises, and practices to manage these systems in a sustainable, resilient manner? What can be managed and at what scales, given that climate change is global in nature but manifests itself at local and regional scales of ecosystems? This section reviews the management models that predict how ecosystems will respond to climate change and examines their adequacy for addressing threshold behavior.

What changes must be made in existing management models, premises, and practices to manage these systems in a sustainable, resilient manner?

5.1. INTEGRATION OF MANAGEMENT AND RESEARCH

With ongoing climate change and the threat that ecosystems will experience threshold changes, managers and decisionmakers are facing more new challenges than ever. Strong partnerships between research and management can help in identifying and providing adaptive management responses to threshold crossings. Because decisionmakers are dealing with whole new ecosystem dynamics, the old ways of managing change do not apply. A new paradigm in which research and management work closely together is needed. The following sections highlight some of the needs of managers.

5.1.1. Need for Conceptual Models

Most frameworks for nonlinear ecosystem behavior are hierarchical so a small number of structuring processes control ecosystem dynamics; each process operates at its own temporal and spatial scale (O'Neill *et al.*, 1986). Finer scales provide the mechanistic understanding for behavior at a particular scale, and broader scales provide the constraints or boundaries on that behavior. Functional relationships between pattern and process are consistent within each domain of scale so that linear extrapolation is possible within a domain (Wiens, 1989). Thresholds occur when pattern-and-process relationships change rapidly with a small or large change in a pattern or environmental driver (Bestelmeyer, 2006; Groffman *et al.*, 2006), although both external stochastic events and internal dynamics can drive systems across thresholds (Scheffer *et al.*, 2001). Crossing a threshold can result in a regime shift where there is a change in the direction of the system and the creation of an alternative stable state (Allen and Breshears, 1998; Davenport *et al.*, 1998; Walker and Meyers, 2004). Under some conditions, thresholds may be recognized when changes in the rate of fine-scale processes within a defined area propagate to produce broad-scale responses (Gunderson and Holling, 2002; Redman and Kinzig, 2003). In these cases, fine-scale processes interact with processes at broader scales to determine system dynamics. A series of cascading thresholds can be recognized where crossing one pattern-and-process threshold induces the crossing of additional thresholds as processes interact (Kinzig *et al.*, 2006). Conceptual models are particularly useful in linking hierarchical models across scales, because the existence of cross-scale interactions are often clearly recognized and can be incorporated as rules, even if they cannot be precisely parameterized. Field experiments that identify cause-and-effect relationships can then be implemented to test these cross-scale interactions. For example, manipulation of CO_2 or water table depth (global-to-regional drivers of change) can be used to assess impacts on plot-scale patterns of biogeochemistry of community composition.

5.1.2. Scaling

Recent theories and ideas about system behavior have used hierarchy theory as a basis for describing interactions among processes at different scales. Such theories include complex systems (Milne, 1998; Allen and Holling, 2002), self-organization (Rietkerk *et al.*, 2004), panarchy (Gunderson and Holling, 2002), and resilience (Holling, 1992; Walker *et al.*, 2006). Cross-scale interactions (CSIs, processes at one spatial or temporal scale interacting with processes at another scale that often result in nonlinear dynamics with thresholds) are an integral part of all of these ideas (Carpenter and Turner, 2000; Gunderson and Holling, 2002; Peters *et al.*, 2004). These interactions generate emergent behavior that cannot be predicted based on observations at single or multiple, independent scales (Michener *et al.*, 2001). CSIs can be important for extrapolating information about fine-scale processes to broad-scales or for downscaling the effects of broad-scale drivers on fine-scale patterns (Ludwig *et al.*, 2000; Diffenbaugh *et al.*, 2005). The relative importance of fine- or broad-scale pattern-and-process relationships can vary through time and compete as the dominant factors controlling system dynamics (for example, Rodó and Comín, 2002; King et al, 2004; Yao *et al.*, 2006).

Because CSI-driven dynamics are believed to occur in a variety of systems, including lotic invertebrate communities in freshwater streams (Palmer *et al.*, 1996) and lakes (Stoffels *et al.*, 2005), mouse populations in forests (Tallmon *et al.*, 2003), soil microbial communities (Smithwick *et al.* 2005), coral reef fish recruitment in the ocean (Cowen *et al.*, 2006), human diseases (Rodó and Comín, 2002), and grass-shrub interactions in deserts (Peters *et al.*,

2006)—it is critical that ecologists find ways to measure CSI. It is important to identify the key processes involved in these changing pattern-and-process relationships so that thresholds can, at a minimum, be understood and predicted if not averted through proactive measures.

Recently, a framework was developed to explain how patterns and processes at different scales interact to create nonlinear dynamics (Peters *et al.*, 2007). This framework focuses on intermediate-scale properties of transfer processes and spatial heterogeneity to determine how pattern-and-process relationships interact from fine to broad scales (fig. 5.1). In this framework, within a domain of scale (that is, fine, intermediate, or broad), patterns and processes can reinforce one another and be relatively stable. Changes in external drivers or disturbances can alter pattern-and-process relationships in two ways.

First, altered patterns at fine scales can result in positive feedbacks that change patterns to the point that new processes and feedbacks are induced. This shift is manifested in a nonlinear threshold change in pattern and process rates. For example, in arid systems, disturbance to grass patches via heavy livestock grazing can reduce the competitive ability of grasses and allow shrub

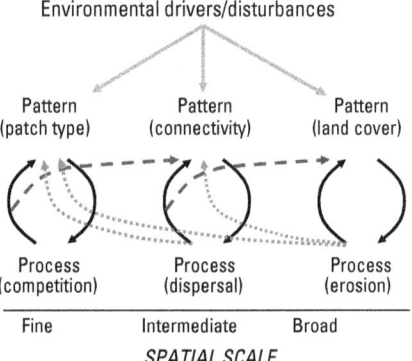

Figure 5.1. Diagram representing cross-scale interactions. Solid arrows represent pattern-and-process feedbacks within three different scale domains with one example of pattern and process shown for each domain. Green arrows indicate the direct effects of environmental drivers or disturbances on patterns or processes at different scales (for example, patch disturbance versus climate). Blue arrows indicate the point at which altered feedbacks at finer scales induce changes in feedbacks at broader scales (for example, fine-scale changes cascade to broader scales). Red arrows indicate when changes at broader scales overwhelm pattern-and-process relationships at finer scales.

A series of cascading thresholds can be recognized where crossing one pattern-and-process threshold induces the crossing of additional thresholds as processes interact.

colonization. After a certain density of shrubs is reached in an area and vectors of propagule transport (for example, livestock or small animals) are available to spread shrubs to nearby grasslands, shrub colonization and grass loss can become controlled by dispersal processes rather than by competition. Shrub expansion rates can increase dramatically (Peters *et al.*, 2006). As shrub colonization and grazing diminish grass cover over large areas, broad-scale wind erosion may govern subsequent losses of grasses and increases in shrub dominance. These broad-scale feedbacks downscale to overwhelm fine-scale processes in remnant grasslands. Once erosion becomes a pervasive landscape-scale process, neither competition nor dispersal effects have significant effects on grass cover.

Second, direct environmental effects on pattern-and-process relationships at broad scales can similarly overwhelm fine-scale processes. For example, regional, long-term drought can produce widespread erosion and minimize the importance of local grass cover or shrub dispersal to patterns in grasses and shrubs.

5.1.3. Applying Models
From Other Disciplines
Climate requires interdisciplinary approaches. Recent and global environmental changes, including climatic change, changes in atmospheric composition, land-use change, habitat fragmentation, pollution, and the spread of invasive species, have the potential to affect the structure and functions of some ecosystems, and the services they provide. Many ecological effects of global environmental change have the potential for feedbacks (either positive or negative) to climatic and other environmental changes. Furthermore, because many global environmental changes are expected to increase in magnitude in the coming decades, the potential exists for more significant effects on ecosystems and their services.

As climate change manifests itself at local and regional scales of ecosystems, it is necessary not only to downscale forecasting models but also to ensure that models used for predictions take into account not just the physical parameters that support ecosystems but also the biotic aspects of the ecosystems. Biomes and ecosystems do not shift as entities in response to climate change, but they change through the responses of individual species (Scott and Lemieux, 2005). The

biogeochemical, temperature, and precipitation requirements of individual species need to be taken into account when predicting these shifts, thus the need for the use of interdisciplinary models that address these variables and their dynamic feedback. Our current understanding suggests that using interdisciplinary models will very likely reduce scientific uncertainties about the potential effects of global change on ecosystems and provide new information on the effects of feedbacks from ecosystems on global change processes. The challenge is to create a framework in which interdisciplinary models can work interactively to consider all the feedbacks involved. This integrative approach will provide a framework to organize observations and assessments of changes in the system in response to management actions.

5.2. ADAPTIVE MANAGEMENT TO INCREASE RESILIENCE

The process of selecting, implementing, monitoring, assessing, and adjusting management actions is called adaptive management or, in the context of this report, adaptive ecosystem management (AEM) (Holling, 1978; Walters, 1986; Prato, 2004; Prato and Fagre, 2007). AEM can be done passively or actively. If passive AEM is used, the decision to adjust management actions or not depends on whether the indicators or multiple attributes of the outcomes of management actions suggest that the ecosystem is becoming more resilient or more variable and might cross a threshold. If active AEM is used, the decision of whether or not to adjust management actions is determined by testing hypotheses about how the ecosystem state is responding to management actions. Active AEM treats management actions as experiments. Unlike passive adaptive management, active AEM yields statistically reliable information about ecosystem responses to management actions, although it is more expensive and difficult to apply than passive AEM and requires sufficient monitoring (Lee, 1993; Wilhere, 2002).

To increase ecosystem resilience, a number of approaches have been put forth for use in adaptive management (Julius *et al.*, 2008). These include avoiding landscape fragmentation and its converse, restoring connectivity; ensuring that refugia are protected so that recolonization of species is possible; focusing protection on keystone species where applicable; reducing other stressors such as pollution; removing introduced invasive species; and reducing extraction of ecosystem services for humans (for example, ensuring water flows for aquatic ecosystems under stress) (Scott and Lemieux, 2005; Groffman *et al.*, 2006). For each ecosystem, AEM potentially provides quantitative documentation as to the relative efficacy of the different approaches to improving resilience (Keeley, 2006; Millar *et al.*, 2007: Newmaster *et al.*, 2007).

5.2.1. Role of Monitoring
Because climate change effects are likely to interact with patterns and processes across spatial and temporal scales, it is clear the monitoring strategies must be integrated across scales. First and foremost, the earth's surface must be hierarchically stratified (for example, using the Major Land Resource Area and Ecological Site Description System of the U.S. Department of Agriculture and National Resources Conservation Service and U.S. Forest Service ecoregions), and conceptual or simulation models of possible impacts and feedbacks must be specified for each stratum (Herrick *et al.*, 2006). The models are used to develop scenarios and to identify key properties and processes that are likely to be associated with abrupt changes. Second, simultaneous multiple-scale monitoring should be implemented at up to three spatial scales based on these scenarios and the recognition of pattern-and-process coupling developed in the models (Bestelmeyer, 2006), which may feature cross-scale interactions (Peters *et al.*, 2004).

Remote-sensing platforms can be used to monitor some broad-scale spatial patterns, including significant shifts in plant community composition; vegetation production; changes in plant mortality; bare-ground, soil, and water-surface temperatures; and water clarity. These platforms may also be used to detect rates of change in some contagious processes, such as the spread of readily observable invasive species. Changes in variance across space and time derived from such measures may be a primary indicator of incipient nonlinear change (Carpenter and Brock, 2004). These measures should be coupled with ground-based measures at mesoscale to patch scales. Mesoscale monitoring often requires widely distributed observations across a landscape

(or ocean) acquired with rapid methodologies including sensor networks. Such widely distributed monitoring is necessary in some situations because incipient changes may materialize in locations that are difficult to predict in advance (such as with tsunami warning systems). In other cases, however, more targeted monitoring is necessary to detect mesoscale discontinuities in smaller areas that are likely to first register broad-scale change, such as at ecotone boundaries (Neilson, 1993). Finally, patch-scale monitoring can feature methodologies that focus on pattern-and-process linkages that scale up to produce systemwide threshold changes, such as when vegetation patches degrade and bare patches coalesce to result in desertification (Rietkerk *et al.*, 2004; Ludwig *et al.*, 2005). The involvement of land users is particularly important at this scale because recognition of processes that degrade resilience may be used to mitigate climate-driven thresholds by way of local management decisions. Consequently, technically sophisticated approaches should be balanced with techniques suitable for the public at large (for example, Carpenter *et al.*, 1999; Pyke *et al.*, 2002).

Monitoring data across scales must then be integrated, and interpretations generated for key strata. Ground-based monitoring, for example, may reveal key changes not detected through remote sensing, or conversely, remote sensing may explain apparently idiosyncratic patterns in ground-based data to reveal key vulnerabilities. Multiagency institutions and a "network of networks" could be organized with such efforts in mind and could periodically review data gathered across scales and from different partners (Parr *et al.*, 2003; Betancourt *et al.*, 2007; Peters *et al.*, 2008).

Monitoring key ecosystem indicators which integrate across ecosystem processes and scales are essential in developing observations for threshold changes. For example, nutrient export via streamflow is a sensitive metric for identifying changes in ecosystem structure and function at the watershed scale that may be difficult to detect on complex and spatially heterogeneous systems. For example, nitrate concentration in streams has been used as a sensitive indicator of forest nitrogen saturation (Stoddard, 1994; Swank and Vose, 1997; Lovett *et al.*, 2000; Aber *et al.*, 2003), effects

Monitoring key ecosystem indicators which integrate across ecosystem processes and scales are essential in developing observations for threshold changes.

of insect pest outbreaks (Eshleman *et al.* 1998), and effects of short-term climate perturbations (Mitchell *et al.*, 1996; Aber *et al.*, 2002). Stream chemistry monitoring, particularly at gauged sites where discharge is also monitored, can provide sensitive signals of changes in ecosystem biogeochemical cycles.

5.2.2. Role of Experiments

A key component of adaptive management strategies is the role of experimentation. A critical component of designing appropriate experiments is to identify the conditions or systems that are susceptible to threshold behavior and interactions across scales that include transport processes at intermediate scales. Observations and experiments to evaluate the sensitivity of these processes and interconnections to anticipated perturbations provide insight to management strategies to enhance resilience or to mitigate threshold changes. One approach is to measure responses at multiple scales simultaneously and then test for significant effects of variables at each scale (for example, Smithwick *et al.*, 2005; Stoffels *et al.*, 2005). Experimental manipulations can also be used to examine processes at fine and intermediate scales and to isolate and measure impacts of broad-scale drivers under controlled conditions (for example, Palmer *et al.*, 1996; King *et al.*, 2004). Stratified-cluster experimental designs are methods for considering multiple scales in spatial variables and for accounting for distance as related to transport processes in the design (Fortin *et al.*, 1989; King *et al.*, 2004). Regression (gradient)-based experimental designs may be superior to analysis of variance (ANOVA)-type designs for predicting thresholds in ecological response to linear or gradual changes in climate or other drivers.

Quantitative approaches also show promise in identifying key processes related to threshold behavior. Statistical analyses based on nonstationarity (Rodó and Comín, 2002) and nonlinear time series analysis (Pascual and Ellner, 2000) are useful for identifying key processes at different scales. Spatial analyses that combine traditional data layers for fine- and broad-scale patterns with data layers that use surrogates for transfer processes at intermediate scales (for example, seed dispersal) can isolate individual processes and combinations of processes that influence dynamics in both space and time (for example, Yao *et al.*, 2006). Simulation

models that use fine-scale models to inform a broad-scale model can be used to examine the relative importance of processes and drivers at different scales to system dynamics as well as interactions of processes and drivers (Moorcroft *et al.*, 2001; Urban, 2005). Coupled biological and physical models that include population processes and connectivity among populations as well as broad-scale drivers have been used to show the conditions when connectivity is important, and to identify the locations that are more susceptible or resilient to management decisions (Cowen *et al.*, 2006).

5.3. MANAGEMENT BY COPING

If there is a high potential for abrupt or threshold-type changes in ecosystems in response to climate change, existing management models, premises, and practices must be modified in order to manage these systems in a sustainable, resilient manner (Millar *et al.*, 2007). Existing management paradigms may have some limited value because of the assumption that the future will be similar to the past. This assumption, however, fails to take into account the underlying uncertainty of the trajectories of ecological succession in the face of climate change. Managers can instead take a dynamic approach to natural resource management, emphasizing processes rather than composition, to best maintain, restore, and enhance ecological functions (Walker *et al.*, 2002). The following sections address some of the mechanisms that can be used to plan for future ecosystem resilience and achieve a balance of positive and negative feedbacks (Millar *et al.*, 2007).

5.3.1. Reducing Multiple Stressors

The key to reducing stressors is to identify the factors that influence resilience. In many cases, management practices that increase resilience can be designed from existing knowledge; in other cases, however, it is not clear what management practices will enhance resilience (Millar *et al.*, 2007). For example, connectivity in a fragmented landscape can be restored by creating corridors for species movement between suitable habitat patches (Gustafson, 1998). Alternatively, inadvertent connectivity that has been established and utilized by invasive species can be removed to reduce stress on the native populations remaining.

To potentially mitigate for threshold crossing,

Observations and experiments to evaluate the sensitivity of these processes and interconnections to anticipated perturbations provide insight to management strategies to enhance resilience or to mitigate threshold changes.

it is *likely* that a variety of approaches, including both long-term and short-term strategies based on new information for natural resource management, will need to focus on increasing ecosystem resilience and resistance as well as assisting ecosystems to adapt to the inevitable changes as climates and environments continue to shift (Millar *et al.*, 2007; Parker *et al.*, 2000). Increasing management adaptive capacity is the operative action taken to increase resilience in ecosystems. For instance, increasing water storage capacity can provide a buffer against reaching the trigger point for a drought-induced threshold crossing that would permanently change an arid land ecosystem. The concept of critical loads for organisms is well established but can be productively applied to ecosystems.

Based on gaps in the literature identified through the development process for this assessment (SAP 4.2) and the synthesis team's expertise, tools to analyze and detect nonlinearity and thresholds from monitoring data will need to be developed. Increases in the variance of an important ecosystem metric have been suggested as an early sign of system instability. As negative feedbacks weaken and positive feedbacks strengthen, the likelihood that a threshold will be reached and crossed increases. As identified by the synthesis team in producing this assessment, there is a need for more nonlinear modeling and statistics to be applied to the threshold issue to identify the point at which positive feedbacks dominate.

5.3.2. Triage

Scientific evidence shows that climate change in the 21st century will most likely result in new vegetation successions, water regimes, wildlife habitat and survival conditions, permafrost and surface-ice conditions, coastal erosion and sea-level change, and human responses (Welch, 2005). Triage is a process in which things are ranked in terms of importance or priority. The term environmental or ecological triage has been used to describe the prioritization process used by policymakers and decisionmakers to determine targets and approaches to dealing with resource allocation (for example, health of ecosystems) that are in high demand and rapidly changing. In the planning process, resource managers can address ecological triage under three different priorities: (1) *status quo* or do nothing; (2) reaction after disturbance; or (3) proactive intervention (Holt and Viney 2001). Triage is

a useful tool to prioritize actions, especially in cases where highly valued resources are at stake, conditions are changing rapidly, and decisions are urgent. The approaches to apply after triage are adaptive management, and mitigation and adaptation strategies. Enabling ecosystems to respond to climate change will help to ease the transition from current to future stable and resilient states and to minimize threshold changes (Fitzgerald, 2000; Holt and Viney, 2001; Millar *et al.*, 2007; Millar, in press).

5.3.3. System-Level Planning and Policy

Expanding management to regional levels is also key, because climate change may be pushing ecosystems to regional synchrony. In order to better understand and to manage thresholds, developing a regional perspective may prove to be more effective than using a local perspective. This regional approach would take into account large scale changes in climate regimes while still incorporating local scale resource issues. An example is that wildland fire is synchronously increasing throughout the western United States and could lead to major recruitment events for species such as lodgepole pine or trigger beetle outbreaks at unprecedented scales. These recruitment events could lead to supercohorts that develop with succession following subcontinental-scale disturbance. There is little management precedent for these types of outcomes that are threshold events on a continental scale, even if they are common on local scales.

Adaptive management and structured decision-making will almost certainly be required to deal with increased temperature effects on threshold crossings and the different trajectories of succession that follow in the western United States. Natural systems are out of sync with climate, leading to the greatest potential for new species combinations in many centuries. Therefore, new actions may be considered, such as planting different tree genotypes after large-scale fires, with appropriate followup monitoring to learn from the results.

5.3.4. Capacity Building and Awareness

There is, and will be, an urgent need to adapt where climate-change-induced thresholds are crossed and a new ecosystem state will be a reality for the foreseeable future. Capacity building basically increases the resilience of the socioeconomic system to tolerate different states of natural resources and ecosystem functioning (Scott

The assumption that the future will be similar to the past fails to take into account the underlying uncertainty of the trajectories of ecological succession in the face of climate change.

and Lemieux, 2005). If ecosystems become more variable in providing essential ecosystem services, greater flexibility is needed on the human side. An example is the need to add storage capacity for capturing mountain ecosystem water if a threshold in snow persistence is crossed, leading to smaller and more variable snowpacks. Building stakeholder tolerance for change is part of the adaptation that will be necessary (Scott and Lemieux, 2005).

Adaptation can take many forms. Scenario planning provides descriptions of plausible future conditions. Scenario planning, done at the local level, makes stakeholders aware of the scope of uncertainty, facilitates tolerance for change, and motivates the desire to build capacity to better handle threshold changes. Multiscenario approaches used with ecosystem modeling can also be used to develop a range of possible post-threshold conditions to better inform strategic decisionmaking and planning for natural resource managers (Lemieux and Scott, 2005). Impact assessments on specific resources (for example, population viability of individual species) can be expanded to examine the underlying viability of protected areas designed to maintain ecosystems (Scott and Suffling, 2000). These assessments can prepare managers by broadening the scope of planning and ensuring that institutional action plans remain flexible.

5.4. SUMMARY

As this synthesis makes clear, climate change increases the likelihood that ecosystems will undergo threshold changes. The underlying mix of interacting feedback mechanisms that drive these thresholds is poorly understood. Monitoring of ecosystems to detect early indicators, such as increasing variability in system behavior, is generally inadequate even when it is known what aspect of the system to monitor. Based on gaps in the literature identified by the synthesis team, there is little scientific or natural resource management experience in dealing with ecosystems undergoing threshold changes. The degree to which we can reverse a threshold change is largely unknown. These knowledge gaps present scientists and resource managers with severe challenges in anticipating and coping with threshold changes to the natural systems.

The gaps identified include the need to increase the resilience of ecosystems and reduce multiple stressors to avoid threshold crossing. Both of these challenges are difficult to plan for but also are consistent with managing ecosystems under conditions of uncertainty such as climate change. After a threshold crossing occurs, viable options are to increase coping mechanisms, adaptive capacity, and stakeholder tolerance. The publication of this assessment (SAP 4.2) will bring the state of scientific understanding to the forefront of the natural resource management paradigm, identifying a need for greater scientific research on thresholds and ecosystem response to adequately manage natural resources for the future.

A variety of approaches, including both long-term and short-term strategies based on new information for natural resource management, will need to focus on increasing ecosystem resilience and resistance as well as assisting ecosystems to adapt to the inevitable changes as climates and environments continue to shift.

CHAPTER 6

Summary and Science Recommendations

Lead Authors: Daniel B. Fagre, USGS; Colleen W. Charles, USGS.
Authors: Craig D. Allen, USGS; Charles Birkeland, USGS-University of Hawaii; F. Stuart Chapin, III, University of Alaska; Peter M. Groffman, Institute of Ecosystem Studies; Glenn R. Guntenspergen, USGS; Alan K. Knapp, Colorado State University; A. David McGuire, USGS-University of Alaska; Patrick J. Mulholland, Oak Ridge National Laboratory; Debra P.C. Peters, USDA Agricultural Research Service; Daniel D. Roby, USGS-Oregon State University; George Sugihara, Scripps Institute of Oceanography and University of California at San Diego.
Contributing Authors: Brandon Bestelmeyer, Jornada Basin LTER; Julio L. Betancourt, USGS; Jeffrey E. Herrick, Jornada Basin LTER

6.1. SUMMARY

Because of the enormous role they are believed to play in the tolerance of ecosystems to climate change, the existence of thresholds should be a key concern of scientists, Federal land managers, and other natural resource professionals responsible for the state of natural resources and the ecological services these resources provide. Sudden large-scale changes in ecosystems may present new challenges to resource managers because the capacity to predict, manage, and adapt to threshold crossings is currently limited. One goal of resource management is to minimize the risk of declines and uncertainty in the delivery of ecological goods and services but, as discussed in chapter 3, thresholds can precipitate such sudden declines and greatly increase management risks. Indeed, efforts by resource managers to reduce variance in the production of particular goods and services lead to a reduction in ecosystem resilience and increase the probability of threshold change. Current regulatory and legal frameworks do not account for threshold behavior of ecosystems. For this reason and because the social and economic costs of these precipitous collapses are potentially high (for example, the collapse of Atlantic cod population), we recommend the following possible actions be considered as a national priority.

> Sudden large-scale changes in ecosystems may present new challenges to resource managers because the capacity to predict, manage, and adapt to threshold crossings is currently limited.

6.2. SCIENCE RECOMMENDATIONS

Given the knowledge that ecological thresholds exist and the lack of tools to predict them, scientists need to develop better predictive capabilities, and managers must make adjustments to increase their capacity to cope with surprises. If climate change is pushing more ecosystems toward thresholds, what can be done at the national level? In the development of SAP 4.2 the following potential actions were identified. The actions (or approaches) are organized according to those that can be taken before, during, and after thresholds of ecological change are crossed.

6.2.1. Before

Support Research To Identify Thresholds.—Although the existence of thresholds of ecological change is widely acknowledged, further advancement and agreement on the nature and effects of thresholds is limited by the small number of empirical studies that address this topic. Further advancement will depend on the development and use of rigorous tests to identify thresholds reliably across different systems.

Enhance Adaptive Capacity.—Given that threshold changes are increasingly likely to occur, a "no-regrets" policy to prepare for them would enhance the capacity of the socioecological system to cope with change—that is, it would increase its resilience. To implement management changes that could reduce the likelihood of threshold changes, resource managers must first determine the factors that influence the resilience of the systems they manage. These determinations should consider the importance both of ecological diversity at patch and landscape scales and of economic diversity and innovation. The key components of diversity and adaptive capacity and resilience would need to be determined on a system-by-system basis and should include consideration of soil, plant, and animal disturbance, socioecological factors, and cross-scale interactions. A key assumption is that management plans that minimize diversity to maximize the provision of one particular ecosystem good or service are likely to increase the susceptibility of the system to threshold changes.

Monitor and Adjust Multiple Factors and Drivers.—Once the key factors that control the adaptive capacity and resilience of a system have been identified, monitoring programs may be altered to include these factors, as well as the resources and ecological services of management interest. For example, monitoring the effects of increased salinity and (or) inundation from sea level rise on vegetation in coastal wetlands may make it possible to predict what degree of stress vegetation can endure before it goes beyond the ability to recover (Burkett *et al.*, 2005). Monitoring soil conditions in areas that are susceptible to nonnative species invasions may make it possible to predict when invasive species may appear in a stressed ecosystem and push it beyond its threshold. It might also be useful to monitor the variability rather than mean values of an ecological service, because an increase in the amplitude of variability is sometimes an indication of system instability before a threshold is crossed. Another potential indicator is a slowing in response time (recovery time) to local perturbations; in certain theoretical scenarios, perturbations may grow larger in amplitude with an ever-increasing period of recovery as a threshold is approached (Van Nes and Scheffer, 2007).

Current understanding suggests that thresholds are likely to be triggered when resource use pressures interact with gradual changes in climate that are associated with extreme climatic events, such as extended drought periods or hurricanes. Adjusting resource use provides one of the few near-term means available to mitigate thresholds. To enable rapid adjustments in resource use in at-risk places and time periods, it would be useful to put in place finer-grained climate and ecosystem monitoring systems coupled with administrative mechanisms to expedite policy modifications.

Develop Scenarios To Explore Alternative Management Options for Dealing With Potential Changes.—The types of changes that cause threshold changes often are well known in advance (for example, hurricanes, wildfire, or invasive species). Scenario analysis with well-characterized dynamics can explore the potential consequences of taking actions either to reduce the likelihood of threshold change or minimize the impact of changes that occur. In this way, scenarios can provide managers with tools for action before the crisis occurs.

Collate and Integrate Information Better at Different Scales.—Cross-scale interaction, where change in a large-scale variable, such as climate, alters a local-scale driver of threshold change, such as fire, is a great challenge in assessing and preventing threshold change. Greater efficiency and use of information is likely to result from coordinating and pooling information from adjoining jurisdictions and different agencies. For example, trends that are not significant or noticeable at small scales may be clear at larger scales. These and other observations argue for much better integration and coordination of monitoring information, not necessarily more monitoring. Although considerable investment would be needed to make monitoring "smarter" initially, the payoff would be the ability to detect early indicators of ecosystem change that could result in a threshold crossing.

Reduce Other Stressors.—The points that may trigger an abrupt change in an ecosystem that is responding to climate change are rarely known because human civilizations have not witnessed climate change of this magnitude. However, the likelihood of crossing a threshold is most likely lessened by reducing other stressors on the ecosystem (Scott and Lemieux, 2005; Julius *et al.*, 2008). These other stressors might include air and water pollution, regional landscape fragmentation, and control of invasive plants. To help reduce stressors, decisions could be made to allow larger or more extensive buffers when considering carrying capacity of habitats, minimum habitat sizes for species of interest, or use of ecological services, such as water.

6.2.2. During

Manage Threshold Shifts.—There may be constraints to reducing or reversing climate-change-induced stresses to components of an

Given the knowledge that ecological thresholds exist and the lack of tools to predict them, scientists need to develop better predictive capabilities, and managers must make adjustments to increase their capacity to cope with surprises.

ecosystem. If a threshold seems likely to occur but the uncertainties remain high as to when it will occur, contingency plans can be created (Julius *et al.*, 2008). These plans can be implemented when the threshold shift begins to occur or they can be carried out in advance if the onset of the threshold crossing is imminent. Take, for example, an Alpine area in which trees have begun to grow at higher elevations than the current tree line because reduced snowpack has lengthened the growing season. If this tree invasion of formerly open areas reduces animal movement between adjoining mountain areas, movement corridors can be kept open by mechanical clearing of trees.

Project Impacts to Natural Resources.—Many efforts are underway to project climate change (for example, Global Climate Models) and ecosystem responses to climate change (for example, mapped atmosphere-plant-soil systems) using simulation models and other tools. These models generally project ecosystem trends and shifts, but they do not explicitly consider the possibility of thresholds as part of the system dynamics. To project impacts to natural resources accurately, it is necessary to understand, model, and project ecosystem responses to climate change with explicit acknowledgment of thresholds. An example of how the inclusion of thresholds in modeling would be beneficial is the bark beetle outbreak now occurring in Western forests where one threshold was passed when warmer winters allowed two lifecycles of beetle reproduction per year rather than one and where a second threshold may be passed by the expansion of the forests northward to connect with boreal forests that provide a corridor eastward. Such a scenario could lead to continental-scale beetle infestation (Logan *et al.*, 1998).

Recognize Need for Decisionmaking at Multiple Scales.—Climate change often expresses itself across regional boundaries which transcend local jurisdiction and management boundaries. The scale of some threshold crossings, such as the bark beetle example above, is likely to require coordinated decisions on larger scales than in the past. Because of different agency management mandates, levels of resources, or geographic scope, the potential exists for agencies to work at cross-purposes when coping with threshold effects at large scales. Also, the effectiveness of response can be enhanced through economies of scale if several agencies work on the problem simultaneously.

Instigate Institutional Change To Increase Adaptive Capacity.—The capacity for synthesis is a critical component of identifying potential thresholds in ecosystem processes on multiple scales. Institutional changes that promote greater interdisciplinary and interagency scientific and information exchange are likely to increase adaptive capacity in general. Such institutional changes would be especially helpful when implementing comprehensive monitoring to detect and document responses to thresholds in ecosystems.

Identify Research Needs and Priorities To Address Thresholds.—Identifying research needs in general can help when evaluating calls for specific threshold research. The ubiquity of threshold problems across so many fields suggests the possibility of finding common principles at work. The cross-cutting nature of the problem of large-scale system change suggests an unusual opportunity to leverage effort from other fields and apply it to investigating the systemic risk of crossing thresholds. Ecological and economic systems share common elements as complex adaptive systems. To the extent that the analogy holds, these two disciplines have potential for mutual leverage. Beyond the specific analogy between ecology and economics, certain dynamic behaviors and structural (topological network) constraints are common to broad classes of systems. Leverage can also occur by sharing methods across disciplines. Such diverse fields as engineering risk analysis, epidemiology, and ecology employ similar methods and research styles. The aim is not to replace conventional approaches but to explore complementary approaches. Exploiting commonalities is one way that leverage is achieved.

As a further reality check on investments in research and development, management agencies can expand on efforts to examine their bottom-line performance as a normal part of the feedback and evaluation process. For many agencies, this will involve evaluating actual forecast skill as a measure of merit, rather than post-hoc fitting and correlation (the products of which may fit an existing paradigm but lack any predictive skill). Obtaining ground truth on this level can validate whether classical management concepts, such as maximum sustained

Current understanding suggests that thresholds are likely to be triggered when resource use pressures interact with gradual changes in climate that are associated with extreme climatic events, such as extended drought periods or hurricanes.

yield in fisheries and other equilibrium concepts and models, are sufficiently useful to be predictive. A periodic evaluation process based on actual (real time) predictive power should indicate whether the model paradigm currently in use is an adequate representation of real systems, and whether the current direction of investments in research and development are on track. This level of verification is essential for effective management of threshold transitions.

6.2.3. After

Although many of the management responses to thresholds should be continued after thresholds have been crossed (for example, monitoring and building ecosystem resilience), human society will largely be faced with adjusting to different ecosystems. These adaptations may be expensive, requiring significant new physical and administrative infrastructures. Capacity building, scenario planning, and adaptive management must all be applied to quickly improve the ability of management to cope with a different ecosystem and for stakeholders to adjust their expectations of ecosystem services.

6.3. CONCLUSION

There is a need to develop a deeper understanding of thresholds of ecological change, especially given our current relative inability to predict when and where they will occur. There have, however, been enough occurrences with significant economic and social costs to warrant consideration of thresholds in natural resource planning and management. Threshold threats to many ecosystems are threats to long-term sustainability of human users as well as biodiversity and biological adaptive capacity. This document has summarized much of what is known about thresholds and has suggested approaches to improve understanding of thresholds, to reduce the chances of threshold crossing, and to enhance the ability to cope with thresholds that have occurred. Given the magnitude of climate change effects on ecosystems, the added factor of sudden threshold changes complicates societal responses and underscores the importance of continued integration of research and management to develop appropriate strategies for coping with thresholds.

GLOSSARY

Adaptive capacity
the capacity of organisms, both individuals and groups, to respond to and change in the state of the system (Folke *et al.*, 2003; Walker *et al.*, 2004; Adger *et al.*, 2005); depends on initial diversity and the capacity of component organisms to adjust and change

Bioerosion
describes the erosion of hard ocean substrates by living organisms by physical mechanisms such as boring, drilling, rasping, and scraping or by chemical mechanisms for dissolution

Degradation
deterioration of a system to a less desirable state as a result of failure to actively adapt or transform

Degree Heating Week
the NOAA satellite-derived Degree Heating Week (DHW) is an experimental product designed to indicate the accumulated thermal stress that coral reefs experience. A DHW is equivalent to 1 week of sea surface temperature 1°C above the expected summertime maximum. For example, 2 DHWs indicate 1 week of 2°C above the expected summertime maximum

Ecosystem
all the organisms, including people, in an area and the nonbiological materials, such as water and soil minerals, with which they interact

Ecosystem services
benefits that people derive from ecosystems, including supporting, provisioning, regulating, and cultural services

Exogenous factor
factor external to the system being managed and which therefore is not incorporated into the management framework

Exposure
nature and degree to which the system experiences environmental or sociopolitical stress

Mitigation
reduction in the exposure of a system to a stress or hazard

Negative feedbacks
interaction in which the effects of two interacting components on one another have opposite signs; generally buffer against changes in the system; an important mechanism enhancing resilience

Positive feedback
interaction in which the effects of two interacting components on one another have the same sign (both positive or both negative); tend to amplify changes in the system, leading to threshold changes in the system

Regime shift
sudden shifts in biota that are driven by ocean climate events

Resilience
capacity of a socioecological system to absorb a spectrum of shocks or perturbations and continue to develop with similar fundamental function, structure, identity, and feedbacks, that is, to remain within a given stability domain (Holling, 1973; Gunderson and Holling, 2002; Walker *et al.*, 2004; Folke, 2006); includes adaptive capacity but also depends on legacies (for example, seed banks) and strong negative feedbacks that might balance positive feedbacks that might destabilize the system

Socioecological system
system in which human activities depend on resources and services provided by ecosystems and ecosystem organization is influenced, to varying degrees, by human activities

Steady state
condition of a system in which there is no net change in system structure or functioning over the time scale of study

Sustainability
use of the environment and resources to meet the needs of the present without compromising the ability of future generations to meet their own needs

Threshold
as defined in this assessment, an ecological threshold is the point at which there is an abrupt

change in an ecosystem quality, property, or phenomenon, or where small changes in one or more external conditions produce large and persistent responses in an ecosystem

Vulnerability
the degree to which a system is likely to experience harm due to exposure to a specified hazard or stress (Turner *et al.*, 2003; Adger, 2006)

REFERENCES

Aber, J.D., S.V. Ollinger, C.T. Driscoll, G.E. Likens, R.T. Holmes, R.J. Freuder, and C.L. Goodale, 2002. Inorganic nitrogen losses from a forested ecosystem in response to physical, chemical, biotic, and climatic perturbations. *Ecosystems,* **5,** 648–658.

Aber, J.D., C.L. Goodale, S.V. Ollinger, M.-L. Smith, A.H. Magill, M.E. Martin, R.A. Hallett, and J.L. Stoddard, 2003. Is nitrogen deposition altering the nitrogen status of northeastern forests? *BioScience,* **53,** 375–389.

Abrego, David, K.E. Ulstrup, B.L. Willis, and M.J.H. van Oppen, 2008. Species-specific interactions between algal endosymbionts and coral hosts define their bleaching response to heat and light stress. *Proceedings of the Royal Society,* **275(1648),** 2273–2282.

Adger, W.N., 2006. Vulnerability. *Global Environmental Change,* **16(3),** 268–281.

Adger, W.N., N.W. Arnell, and E.L. Tompkins, 2005. Successful adaptation to climate change across scales. *Global Environmental Change,* **15,** 77–86.

Albertson, F.W., and J.E. Weaver, 1942. History of the native vegetation of western Kansas during seven years of continuous drought. *Ecological Monographs,* **12(1),** 23–51.

Albertson, F.W., and J.E. Weaver, 1945. Injury and death or recovery of trees in prairie climate. *Ecological Monographs,* **15(4),** 393–433.

Allan, J.D., 2004. Landscapes and riverscapes: The influence of land use on stream ecosystems. *Annual Review* of *Ecology, Evolution, and Systematics,* **35,** 257–284.

Allen, C.D., 2007. Interactions across spatial scales among forest dieback, fire, and erosion in northern New Mexico landscapes. *Ecosystems,* **10(5),** 797–808. doi:10.1007/s10021-007-9057-4.

Allen, C.D., and D.D. Breshears, 1998. Drought-induced shift of a forest/woodland ecotone: Rapid landscape response to climate variation. *Proceedings of the National Academy of Sciences,* **95,** 14839–14842.

Allen, C.D., and D.D. Breshears, 2007. Climate-induced forest dieback as an emergent global phenomenon. *Eos, Transactions, AGU,* **88(47),** 504.

Allen, C.R., and C.S. Holling, 2002. Adaptive inference for distinguishing credible from incredible patterns in nature. *Ecosystems,* **5(4),** 319–328.

Anderson, J.T., D. Van Holliday, Rudy Kloser, D.G. Reid, and Yvan Simard, 2008. Acoustic seabed classification: Current practice and future directions. *Journal of Marine Science,* **65(6),** 1004–1011.

Anderson, P.J., and J.F. Piatt, 1999. Community reorganization in the Gulf of Alaska following ocean climate regime shift. *Marine Ecology Progress Series,* **189,** 117–123.

Arft, A.M., M.D. Walker, J. Gurevitch, J.M. Alatalo, M.S. Bret-Harte, M. Dale, M. Diemer, F. Gugerli, G.H.R. Henry, M.H. Jones, R. Hollister, I.S. Jónsdóttir, K. Laine, E. Lévesque, G.M. Marion, U. Molau, P. Mølgaard, Nordenhäll, V. Raszhivin, C.H. Robinson, G. Starr, A. Stenström, M. Stenström, Ø. Totland, L. Turner, L. Walker, P. Webber, J.M. Welker, and P.A. Wookey, 1999. Response patterns of tundra plant species to experimental warming: A meta-analysis of the International Tundra Experiment. *Ecological Monographs,* **69,** 491–511.

Baker, A.C., 2003. Flexibility and specificity in coral-algal symbiosis: Diversity, ecology, and biogeography of *Symbiodinium. Annual Review of Ecology, Evolution, and Systematics,* **34,** 661–689.

Baker, A.C., 2004. Symbiont diversity on coral reefs and its relationship to bleaching resistance and resilience. In: *Coral Health and Disease* [Eugene Rosenberg and Yoshi Loya (eds.)]. Springer-Verlag, Berlin, 177–191.

Baker A.C., C.J. Starger, T.R. McClanahan, and P.W. Glynn, 2004. Coral reefs: Corals' adaptive response to climate change. *Nature*, **430**, 741.

Balshi, M.S., A.D. McGuire, Paul Duffy, Mike Flannigan, John Walsh, and Jerry Melillo, 2008. Assessing the response of area burned in western boreal North America using a multivariate adaptive regression splines (MARS) approach. *Global Change Biology,* doi:10.1111/j.1365-2486.2008.01679.x.

Barber, V.A., G.P. Juday, and E. Berg, 2002. Assessment of recent and possible future forest responses to climate in boreal Alaska. In: *Workshop on Northern Timberline Forests: Environmental and Socio-economic Issues and Concerns.* Finnish Forest Research Institute. Kolari Research Station, 102–105, 288–289.

Barnett, T.P., D.W. Pierce, H.G. Hidalgo, C. Bonfils, B.D. Santer, T. Das, G. Bala, A.W. Wood, T. Nozawa, A.A. Mirin, D.R. Cayan, and M.D. Dettinger, 2008. Human-induced changes in the hydrology of the western United States. *Science*, **319**, 1080–1083.

Barnett, T.P., J.C. Adam, and D.P. Lettenmaier, 2005. Potential impacts of a warming climate on water availability in snow-dominated regions. *Nature*, **438**, 303–309.

Beckage, Brian, Ben Osborne, D.G. Gavin, Carolyn Pucko, Thomas Siccama, and Timonthy Perkins, 2008. A rapid upward shift of a forest ecotone during 40 years of warming in the Green Mountains of Vermont. *Proceedings of the National Academy of Sciences,* **105**, 4197–4202.

Berkelmans, R., and M.J.H. van Oppen, 2006. The role of zooxanthellae in the thermal tolerance of corals: A "nugget of hope" for coral reefs in an era of climate change. *Proceedings of the Royal Society of London, Series B: Biological Sciences,* **273**, 2305–2312.

Bestelmeyer, B.T, 2006. Threshold concepts and their use in rangeland management and restoration: The good, the bad, and the insidious. *Restoration Ecology,* **14(3)**, 325–329.

Betancourt, J.L., M.D. Schwartz, D.D. Breshears, C.A. Brewer, G. Frazer, J.E. Gross, S.J. Mazer, B.C. Reed, and B.E. Wilson, 2007. Evolving plans for the USA National Phenology Network. *Eos, Transactions, AGU,* **88(19)**, 211.

Bigler, Christof, Dominik Kulakowski, and T.T. Veblen, 2005. Multiple disturbance interactions and drought influence fire severity in Rocky Mountain subalpine forests. *Ecology,* **86(11)**, 3018–3029.

Birkeland, Charles, 2004. Ratcheting down the coral reefs. *BioScience*, **54**, 1021–1027.

Birkeland, Charles, Peter Craig, Douglas Fenner, Lance Smith, W.E. Kiene, and B.M. Riegl, 2008. Geologic setting and ecological functioning of coral reefs in American Samoa. In: *Coral Reefs of the USA* [B.M. Riegl and R.E. Dodge (eds.)]. Springer, **1**, 741–766.

Boulton, A.J., 2003. Parallels and contrasts in the effects of drought on stream macroinvertebrate assemblages. *Freshwater Biology*, **48**, 1173–1185.

Boulton, A.J., C.J. Peterson, N.B. Grimm, and S.G. Fisher, 1992. Stability of an aquatic macroinvertebrate community in a multiyear hydrologic disturbance regime. *Ecology*, **73**, 2192–2207.

Breshears, D.D., 2006. The grassland-forest continuum: Trends in ecosystem properties for woody plant mosaics? *Frontiers in Ecology and the Environment*, **4(2)**, 96–104.

Breshears, D.D., 2007. Drought-induced vegetation mortality and associated ecosystem responses: Examples from semiarid woodlands and forests. In: *Understanding Multiple Environmental Stresses.* National Academy Press, Washington, D.C., 89–95.

Breshears, D.D., and C.D. Allen, 2002. The importance of rapid, disturbance-induced losses in carbon management and sequestration. *Global Ecology and Biogeography Letters*, **11**, 1–15.

Breshears, D.D., and F.J. Barnes, 1999. Interrelationships between plant functional types and soil moisture heterogeneity for semiarid landscapes within the grassland/forest continuum: A unified conceptual model. *Landscape Ecology*, **14**, 465–478.

Breshears, D.D., N.S. Cobb, P.M. Rich, K.P. Price, C.D. Allen, R.G. Balice, W.H. Romme, J.H. Kastens, M.L. Floyd, Jayne Belnap, J.J. Anderson, O.B. Myers, and C.W. Meyer, 2005. Regional vegetation die-off in response to global-change-type drought. *Proceedings of the National Academy of Sciences,* **102**, 15144–15148.

Bruno, J.E., and E.R. Selig, 2007. Regional decline of coral cover in the Indo-Pacific: Timing, extent, and subregional comparisons. *PLoS ONE*, **8**, 711–718.

Buddemeier, R.W., A.C. Baker, D.G. Fautin, and J.R. Jacobs, 2004. The adaptive hypothesis of bleaching. In: *Coral Health and Disease* [Eugene Rosenberg and Yossi Loya (eds.)]. Springer-Verlag, Berlin, 427–444.

Burkett, V.R., D.A. Wilcox, Robert Stottlemyer, Wylie Barrow, Dan Fagre, Jill Baron, Jeff Price, J.L. Nielsen, C.D. Allen, D.L. Peterson, Greg Ruggerone, and Thomas Doyle, 2005. Nonlinear dynamics in ecosystem response to climatic change: Case studies and policy implications. *Ecological Complexity,* **2,** 357–394.

Byrd, G.V., W.J. Sydeman, H.M. Renner, and Shoshiro Minobe, 2008. Contrasting responses of piscivorous seabirds at the Pribilof Islands to ocean climate. *Deep Sea Research Part II: Topical Studies in Oceanography,* **55(16–17),** 1856-1867.

Carpenter, S.R., 2002. Ecological futures: Building an ecology of the long now. *Ecology,* **83,** 2069–2083.

Carpenter, S.R., and W.A. Brock, 2004. Spatial complexity, resilience, and policy diversity: Fishing on lake-rich landscapes. *Ecology and Society,* **9(1),** 8.

Carpenter, S.R., and Monica Turner, 2000. Opening the black boxes: Ecosystem science and economic valuation. *Ecosystems,* **3(1),** 1–3.

Carpenter, Stephen, William Brock, and Paul Hanson, 1999. Ecological and social dynamics in simple models of ecosystem management. *Ecology and Society,* **3(2),** 4. *www.consecol.org/vol3/iss2/art4.*

Chan, F., J.A. Barth, J. Lubchenco, A. Kirincich, H. Weeks, W.T. Peterson, and B.A. Menge, 2008. Emergence of anoxia in the California current large marine ecosystem. *Science,* **319,** 920.

Chapin, F.S., III, G.R. Shaver, A.E. Giblin, K.J. Nadelhoffer, and J.A. Laundre, 1995. Responses of arctic tundra to experimental and observed changes in climate. *Ecology,* **76(3),** 694–711.

Chapin, F.S., III, M.S. Bret-Harte, S.E. Hobbie, and Hailin Zhong, 1996. Plant functional types as predictors of transient responses of arctic vegetation to global change. *Journal of Vegetation Science,* **7(3),** 347–358.

Chapin, F.S., III, M. Sturm, M.C. Serreze, J.P. McFadden, J.R. Key, A.H. Lloyd, A.D. McGuire, T.S. Rupp, A.H. Lynch, J.P. Schimel, J. Beringer, W.L. Chapman, H.E. Epstein, E.S. Euskirchen, L.D. Hinzman, G. Jia, C.-L. Ping, K.D. Tape, C.D.C. Thompson, D.A. Walker, and J.M. Welker, 2005. Role of land-surface changes in arctic summer warming. *Science,* **310,** 657–660.

Chapin, F.S., III, S.F. Trainor, O. Huntington, A.L. Lovecraft, E. Zavaleta, D.C. Natcher, A.D. McGuire, J.L. Nelson, L. Ray, M. Calef, N. Fresco, H. Huntington, T.S. Rupp, L. DeWilde, and R.A. Naylor, 2008. Increasing wildfire in Alaska's boreal forest: Causes, consequences, and pathways to potential solutions of a wicked problem. *BioScience,* **58(6),** 531–540.

Claussen, Martin, Claudia Kubatzki, Victor Brovkin, Andrey Ganopolski, Phillipp Hoelzmann, Hans-Joachim Pachur, 1999. Simulation of an abrupt change in Saharan vegetation in the mid-Holocene. *Geophysical Research Letters,* **26(14),** 2037–2040.

Cohen, Anne, Nicholas Jachowski, Ross Jones, and Struan Smith, 2008. Declining calcification rates of Bermudan brain corals over the past 50 years. In: *11th International Coral Reef Symposium,* Fort Lauderdale, FL, July 7–11, 2008, abstract. National Coral Reef Institute, Fort Lauderdale.

Copper, P., 1994. *Ancient reef ecosystem expansion and collapse.* Coral Reefs, 13, 3–11.

Corn, P.S., 2003. Amphibian breeding and climate change: Importance of snow in the mountains. *Conservation Biology,* **17(2),** 622–625.

Cornelissen, J.H.C., T.V. Callaghan, J.M. Alatalo, A. Michelsen, E. Graglia, A.E. Hartley, D.S. Hik, S.E. Hobbie, M.C. Press, C.H. Robinson, G.H.R. Henry, G.R. Shaver, G.K. Phoenix, D. Gwynn Jones, S. Jonasson, F.S. Chapin III, U. Molau, C. Neill, J.A. Lee, J.M. Melillo, B. Sveinbjörnsson, and R. Aerts, 2001. Global change and arctic ecosystems: Is lichen decline a function of increases in vascular plant biomass? *Journal of Ecology,* **89,** 984–994.

Covich, A.P., S.C. Fritz, P.J. Lamb, R.D. Marzolf, W.J. Matthews, K.A. Poiani, E.E. Prepas, M.B. Richman, and T.C. Winter, 1997. Potential effects of climate change on aquatic ecosystems of the Great Plains of North America. *Hydrological Processes,* **11(8),** 993–1021.

Cowen, R.K., C.B. Paris, and A. Srinivasan, 2006. Scaling of connectivity in marine populations. *Science,* **311,** 522–527.

Coyle, K.O., V.V. Chavtur, and A.I. Pinchuk, 1996. Zooplankton of the Bering Sea. In: *Ecology of the Bering Sea: A Review of Russian Literature* [O.A. Mathisen and K.O. Coyle (eds.)]. Alaska Sea Grant College Program, Fairbanks, 97–133.

Criddle, K.R., M. Herrmann, J.A. Greenberg, and E.M. Feller, 1998. Climate fluctuations and revenue maximization in the eastern Bering Sea fishery for walleye pollock. *North American Journal of Fisheries Management*, **18**, 1–10.

Crutzen, P.J., and J.G. Goldammer, 1993. *Fire in the Environment: The Ecological, Atmospheric, and Climatic Importance of Vegetation Fires.* Dahlem Konferenz (March 15–20, 1992, Berlin), Wiley, Chichester, United Kingdom.

Davenport, D.W., D.D. Breshears, B.P. Wilcox, and C.D. Allen. 1998. Viewpoint: Sustainability of piñon-juniper ecosystems—A unifying perspective of soil erosion thresholds. *Journal of Range Management*, **51(2)**, 229–238.

Denoël, Mathieu, and G.F. Ficetola, 2007. Landscape-level thresholds, and newt conservation. *Ecological Applications*, **17(1)**, 302–309.

Dennis, R.L.H., and T.G. Shreeve, 1991. Climatic change and the British butterfly fauna: Opportunities and constraints. *Biological Conservation*, **55**, 1–16.

Diffenbaugh, N.S., J.S. Pal, R.J. Trapp, and Filippo Giorgi, 2005. Fine-scale processes regulate the response of extreme events to global climate change. *Proceedings of the National Academy of Sciences*, **102**, 15,774–15,778.

Dixon, P.A., M.J. Milicich, and George Sugihara, 1999. Episodic fluctuations in larval supply. *Science*, **283(5407)**, 1528–1530.

Dobbertin, Matthias, Beat Wermelinger, Christof Bigler, Matthias Börgi, Mathias Carron, Beat Forster, Urs Gimmi, and Andreas Rigling, 2007. Linking increasing drought stress to scots pine mortality and bark beetle infestations. Proceedings: Impacts of Air Pollution and Climate Change on Forest Ecosystems. *The Scientific World Journal*, **7(S1)**, 231–239.

Dodds, W.K., Keith Gido, M.R. Whiles, K.M. Fritz, and W.J. Matthews, 2004. Life on the edge: The ecology of Great Plains prairie streams. *BioScience*, **54(3)**, 205–216.

Done, Terry, Lyndon Devantier, Emre Turak, Mary Wakeford, Abbi McDonald, and Craig Johnson, [in press]. Great Barrier Reef coral communities: Resilient in the 1980s but struggling in the 2000s. In: *11th International Coral Reef Symposium*, Fort Lauderdale, FL, July 7–11, 2008, abstract, 106. National Coral Reef Institute, Fort Lauderdale.

Douglas, M.R., P.C. Brunner, and M.E. Douglas, 2003. Drought in an evolutionary context: Molecular variability in flannelmouth sucker (*Catostomus latipinnis*) from the Colorado River basin of western North America. *Freshwater Biology*, **48**, 1254–1273.

Dublin, H.T., A.R.E. Sinclair, J. McGlade, 1990. Elephants and fire as causes of multiple stable states in the Serengeti-Mara woodlands. *The Journal of Animal Ecology*, **59(3)**, 1147–1164.

Dye, D.G., 2002. Variability and trends in the annual snow-cover cycle in Northern Hemisphere land areas, 1972–2000. *Hydrological Processes*, **16,** 3065–3077.

Dye, D.G., and C.J. Tucker, 2003. Seasonality and trends of snow-cover, vegetation index, and temperature in northern Eurasia. *Geophysical Research Letters*, **30**, 45, doi:10.1029/GL016384.

Easterling, D.R., G.A. Meehl, C. Parmesan, S.A. Changnon, T.R. Karl, and L.O. Mearns, 2000. Climate extremes: Observations, modeling, and impacts. *Science,* **289**, 2068–2074.

EPRI, 2003. *A Survey on Water Use and Sustainability in the United States With a Focus on Power Generation.* Electric Power Research Institute, Palo Alto, CA, Topical Report 1005474.

Eshleman, K.N., R.P. Morgan, J.R. Webb, F.A. Deviney, and J.N. Galloway, 1998. Temporal patterns of nitrogen leakage from mid-Appalachian forested watersheds: Role of forest disturbance. *Water Resources Research*, **34**, 2005–2116.

Euskirchen, E.S., A.D. McGuire, D.W. Kicklighter, Q. Zhuang, J.S. Clein, R.J. Dargaville, D.G. Dye, J.S. Kimball, K.C. McDonald, J.M. Melillo, V.E. Romanovsky, and N.V. Smith, 2006. Importance of recent shifts in soil thermal dynamics on growing season length, productivity, and carbon sequestration in terrestrial high-latitude ecosystems. *Global Change Biology*, **12(4)**, 731–750.

Euskirchen, E.S., A.D. McGuire, and F.S. Chapin III, 2007. Energy feedbacks of northern high-latitude ecosystems to the climate system due to reduced snow cover during 20th century warming. *Global Change Biology*, **13(11)**, 2425–2438.

Feely, R.A., C.L. Sabine, J. M. Hernandez-Ayon, Debby Lanson, and Burke Hales, 2008. Evidence for upwelling of corrosive "acidified" water onto the continental shelf. *Science,* **320(5882)**, 1490–1492.

Fensham R.J., and J.E. Holman, 1999. Temporal and spatial patterns in drought-related tree dieback in Australian savanna. *Journal of Applied Ecology,* **36**, 1035–1060.

Fisher, M., 1997. Decline in the juniper woodlands of Raydah reserve in southwestern Saudi Arabia: A response to climate changes? *Global Ecology and Biogeography Letters,* **6**, 379–386.

Fitzgerald, G., 2000. Triage. In: *Textbook of Adult Emergency Medicine* [P. Cameron, G. Jelinke, A. M. Kelly, L. Murray, and J. Heyworth (eds.)]. Churchill Livingston, Sydney, 584–588.

Fleming, R.A., and W.J.A. Volney, 1995. Effects of climate change on insect defoliator population processes in Canada's boreal forest: Some plausible scenarios. *Water, Air, & Soil Pollution,* **82(1–3)**, 445–454.

Foley, J.A., M.T. Coe, Marten Scheffer, and Guiling Wang, 2003. Regime shifts in the Sahara and Sahel: Interactions between ecological and climatic systems in Northern Africa. *Ecosystems,* **6(6)**, 524–532.

Folke, Carl, 2006. Resilience: The emergence of a perspective for social-ecological systems analyses. *Global Environmental Change,* **16(3)**, 253–267.

Folke, Carl, Johan Colding, and Fikret Berkes, 2003. Building resilience and adaptive capacity in social-ecological systems. In: *Navigating Social-Ecological Systems* [Fikret Berkes, Johan Colding, and Carl Folke (eds.)]. Cambridge University Press, 352–387.

Fortin, M.-J., Pierre Drapeau, and Pierre Legendre, 1989. Spatial autocorrelation and sampling design in plant ecology. *Plant Ecology,* **83(1–2)**, 209–222.

Gardner, T.A., I.M. Côté, J.A. Gill, Alastair Grant, and A.R. Watkinson, 2003. Long-term region-wide declines in Caribbean corals. *Science,* **301**, 958–960.

Garrison, V.H., E.A. Shinn, W.T. Foreman, D.W. Griffin, C.W. Holmes, C.A. Kellogg, M.S. Majewski, L.L. Richardson, K.B. Ritchie, and G.W. Smith, 2003. African and Asian dust: From desert soils to coral reefs. *BioScience,* **53**, 469–480.

Gibson, C.A., J.L. Meyer, N.L. Poff, L.E. Hay, and A. Georgakakos, 2005. Flow regime alterations under changing climate in two river basins: Implications for freshwater ecosystems. *River Research and Applications,* **21**, 849–864.

Gitlin, A.R., C.M. Stultz, M.A. Bowker, Stacy Stumpf, K.L. Paxton, Karla Kennedy, Axhel Muñoz, J.K. Bailey, and T.G. Whitham, 2006. Mortality gradients within and among dominant plant populations as barometers of ecosystem change during extreme drought. *Conservation Biology,* **20(5)**, 1477–1486.

Goetz, S.J., A.G. Bunn, G.J. Fiske, and R.A. Houghton, 2005. Satellite-observed photosynthetic trends across boreal North America associated with climate and fire disturbance. *Proceedings of the National Academy of Sciences,* **102(38)**, 13521–13525.

Gonzalez, P., 2001. Desertification and a shift of forest species in the West African Sahel. *Climate Research,* **17**, 217–228.

Grebmeier, J.M., J.E. Overland, S.E. Moore, E.V. Farley, E.C. Carmack, L.W. Cooper, K.E. Frey, J.H. Helle, F.A. McLaughlin, and S.L. McNutt, 2006. A major ecosystem shift in the northern Bering Sea. *Science,* **311**, 1461–1464.

Grenfell, B.T., 1992. Chance and chaos in measles dynamics. *Journal of the Royal Statistical Society, series B (methodological),* **54(2)**, 383–398.

Groffman, P.M., J.S. Baron, Tamara Blett, A.J. Gold, Iris Goodman, L.H. Gunderson, B.M. Levinson, M.A. Palmer, H.W. Paerl, G.D. Peterson, N.L. Poff, D.W. Rejeski, J.F. Reynolds, M.G. Turner, K.C. Weathers, and John Wiens, 2006. Ecological thresholds: The key to successful environmental management or an important concept with no practical application? *Ecosystems,* **9**, 1–13.

Gu, L., P.J. Hanson, W.M. Post, D.P. Kaiser, B. Yang, R. Nemani, S.G. Pallardy, and T. Meyers, 2008. The 2007 Easter U.S. spring freeze: Increased cold damage in a warming world?. *BioScience,* **58**, 253–262.

Gunderson, Lance, and C.S. Holling (eds.), 2002. *Panarchy: Understanding Transformations in Human and Natural Systems.* Island Press, Washington, D.C.

Gustafson, E.J., 1998. Quantifying landscape spatial pattern: What is the state of the art? *Ecosystems,* **1**, 143–156.

Hallock, P., 1997. Reefs and reef limestones in earth history. In: *Life and Death of Coral Reefs* [Charles Birkeland (ed.)]. Chapman and Hall, New York, 13–42.

Hallock, P., F.E. Müller-Karger, and J.C. Halas, 1993. Anthropogenic nutrients and the degradation of Caribbean coral reefs. *National Geographic Research and Explorations,* **9(3)**, 358–378.

Hansell, R.I.C., J.R. Malcolm, Harold Welch, R.L. Jeffries, and P.A. Scott, 1998. Atmospheric change and biodiversity in the Arctic. *Environmental Monitoring and Assessment,* **49(2–3),** 303–325.

Hansen, James, Makiko Sato, Reto Ruedy, Ken Lo, D.W. Lea, and Martin Medina-Elizade, 2006. Global temperature change. *Proceedings of the National Academy of Sciences,* **103(30),** 14288–14293.

Hanski, Ilkka, Peter Turchin, Erkki Korpimaki, and Heikki Henttonen, 1993. Population oscillations of boreal rodents: Regulation by mustelid predators leads to chaos. *Nature,* **364,** 232–235.

Hare, S.R., and N.J. Mantua, 2000. Empirical evidence for North Pacific regime shifts in 1977 and 1989. *Progress in Oceanography,* **47(2–4),** 103–145.

Hatfield, J.L., K.J. Boote, B.A. Kimball, D.W. Wolfe, D.R. Ort, R.C. Izaurralde, A.M. Thomson, J.A. Morgan, H.W. Polley, P.A. Fay, T.L. Mader, G.L. Hahn, 2008. Agriculture. In: *The Effects of Climate Change on Agriculture, Land Resources, Water Resources, and Biodiversity.* Synthesis and assessment product 4.3 by the U.S. climate change science program and the subcommittee on global change research, Washington, D.C.

Herrick, J.E., B.T. Bestelmeyer, S. Archer, A.J. Tugel, and J.R. Brown, 2006. An integrated framework for science-based arid land management. *Journal of Arid Environments,* **65,** 319–335.

Higgins, P. A. T., M. D. Mastrandrea, and S. H. Schneider, 2002. Dynamics of climate and ecosystem coupling: Abrupt changes and multiple equilibria. *Philosophical Transactions of the Royal Society of London Series B– Biological Sciences,* **357,** 647–655.

Hinzman L.D., N.D. Bettez, W.R. Bolton, F.S. Chapin, M.B. Dyurgerov, C.L. Fastie, Brad Griffith, R.D. Hollister, Allen Hope, H.P. Huntington, A.M. Jensen, G.J. Jia, Torre Jorgenson, D.L. Kane, D.R. Klein, Gary Kofinas, A.H. Lynch, A.H. Lloyd, A.D. McGuire, F.E. Nelson, W.C. Oechel, T.E. Osterkamp, C.H. Racine, V.E. Romanovsky, R.S. Stone, D.A. Stow, Matthew Sturm, C.E. Tweedie, G.L. Vourlitis, M.D. Walker, D.A. Walker, P.J. Webber, J.M. Welker, K.S. Winker, and Kenji Yoshikawa, 2005. Evidence and implications of recent climate change in terrestrial regions of the Arctic. *Climatic Change,* **72(3),** 251–298.

Hobbs, R.J., S. Arico, J. Aronson, J.S. Barron, P. Bridgewater, V.A. Cramer, P.R. Epstein, J.J. Ewel, C.A. Klink, A.E. Lugo, D. Norton, D. Ojima, D.M. Richardson, E.W. Sanderson, F. Valladares, M. Vila, R. Zamora, M. Zobel, 2006. Novel ecosystems: Theoretical and management aspects of the new ecological world order. *Global Ecology and Biogeography,* **15,** 1–7.

Hoegh-Guldberg, O., P.J. Mumby, A.J. Hooten, R.S. Steneck, P. Greenfield, E. Gomez, C.D. Harvell, P.F. Sale, A.J. Edwards, K. Caldeira, N. Knowlton, C.M. Eakin, R. Iglesias-Prieto, N. Muthiga, R.H. Bradbury, A. Dubi, and M.E. Hatziolos, 2007. Coral reefs under rapid change and ocean acidification. *Science,* **318(5857),** 1737–1742.

Holland, M., C. Blitz, and B. Tremblay, 2006. Future abrupt reductions in the summer Arctic sea ice. *Geophysical Research Letters,* **33,** L23503.

Holling, C.S., 1973. Resilience and stability of ecological systems. *Annual Review* of *Ecology, Evolution, and Systematics,* **4,** 1–23.

Holling, C.S., 1986. The resilience of terrestrial ecosystems; local surprise and global change. In: *Sustainable Development of the Biosphere* [W.C. Clark and R.E. Munn (eds.)]. Cambridge University Press, 292–317.

Holling, C.S., 1992. Cross-scale morphology, geometry, and dynamics of ecosystems. *Ecological Monographs,* **62(4),** 447–502.

Holling, C.S., ed., 1978. *Adaptive Environmental Assessment and Management.* John Wiley & Sons, London.

Holt, D. and H. Viney, 2001. Targeting environmental improvements through ecological triage. *Eco-Management and Auditing,* **8,** 154–164.

Houghton, J.T., L.G. Meira Filho, B.A. Callander, N. Harris, A. Kattenberg, K. Maskell, 1996. Climate change 1995: The science of climate change. Cambridge University Press.

Hsieh, C.H., C.S. Reiss, J.R. Hunter, J.R. Beddington, R.M. May, and George Sugihara, 2006: Fishing elevates variability in the abundance of exploited species. *Nature,* **443,** 859–862.

Hsieh, Chih-hao, Christian Reiss, William Watson, M.J. Allen, J.R. Hunter, R.N. Lea, R.H. Rosenblatt, P.E. Smith, and George Sugihara, 2005. A comparison of long-term trends and variability in populations of larvae of exploited and unexploited fishes in the Southern California region: A community approach. *Progress in Oceanography,* **67(1–2),** 160–185.

Hubbard, D.K., 1997. Reefs as dynamic systems. In: *Life and Death of Coral Reefs* [C. Birkeland (ed.)]. Chapman and Hall, New York, 43–67.

Huggett, A.J., 2005. The concept and utility of 'ecological thresholds' in biodiversity conservation. *Biological Conservation,* **124(3),** 301–310.

Hunt, G.L., Jr., and P.J. Stabeno, 2002. Climate change and the control of energy flow in the southeastern Bering Sea. *Progress in Oceanography,* **55(1–2),** 5–22.

Hunt, G.L., Jr., Phyllis Stabeno, Gary Walters, Elizabeth Sinclair, R.D. Brodeur, J.M. Napp, and N.A. Bond, 2002. Climate change and the control of the southeastern Bering Sea pelagic ecosystem. *Deep Sea Research II,* **49(26),** 5821–5853.

IPCC, 1996. *Climate Change 1995: The Science of Climate Change.* Cambridge University Press.

IPCC, 2007. *Climate Change 2007 Synthesis Report: Contribution of Working Groups I, II and III to the Fourth Assessment Report of the Intergovernmental Panel on Climate Change* [Core Writing Team, R.K. Pachauri, and Andy Reisinger (eds.)]. Intergovernmental Panel on Climate Change, Geneva, Switzerland, 104.

ISRS, 2007. *Coral Reefs and Ocean Acidification.* International Society for Reef Studies, Briefing Paper 5.

Jia, G.J., H.E. Epstein, and D.A. Walker, 2003. Greening of arctic Alaska, 1981–2001. *Geophysical Research Letters,* **30,** 2067, doi:10.1029/2003GL018268,2003.

Johnson, W.C., B.V. Millett, Tagir Gilmanov, R.A. Voldseth, G.R. Guntenspergen, and D.E. Naugle, 2005. Vulnerability of northern prairie wetlands to climate change. *BioScience,* **55(10),** 863–872.

Jorgenson, M. Torre, C.H. Racine, J.C. Walters, T.E. Osterkamp, 2001. Permafrost degradation and ecological changes associated with a warming climate in central Alaska. *Climatic Change,* **48(4),** 551–579.

Juday, G.P., V. Barber, P. Duffy, H. Linderholm, T.S. Rupp, S. Sparrow, E. Vaganov, and J. Yarie, 2005. Forests, land management, and agriculture (chapter 14). In: *Arctic Climate Impact Assessment,* Cambridge University Press, 781–862.

Julius, S., J. West, J. Baron, L. Joyce, P. Kareiva, B. Keller, M. Palmer, C. Peterson, and J.M. Scott, 2008. Preliminary review of adaptation options for climate-sensitive ecosystems and resources. Synthesis and assessment product 4.4 by the U.S. climate change science program and the subcommittee on global change research, Washington, D.C.

Kasischke, E.S., and M.R. Turetsky, 2006. Recent changes in the fire regime across the North American boreal region—Spatial and temporal patterns of burning across Canada and Alaska. *Geophysical Research Letters,* **33,** L09703, doi:09710.01029/02006GL025677.

Kasischke, E.S., David Williams, and Donald Barry, 2002. Analysis of the patterns of large fires in the boreal forest region of Alaska. *International Journal of Wildland Fire,* **11(2),** 131–144.

Kasischke, E.S., T.S. Rupp, and D.L. Verbyla, 2006. Fire trends in the Alaskan boreal forest. In: *Alaska's Changing Boreal Forest* [F.S. Chapin III, M.W. Oswood, K. Van Cleve, L.A. Viereck, and D.L. Verbyla (eds.)]. Oxford University Press, New York, 285–301.

Keeley, J.E., 2006. Fire management impacts on invasive plants in the western United States. *Conservation Biology,* **20,** 375–384.

King, R.S., C.J. Richardson, D.L. Urban, and E.A. Romanowicz, 2004. Spatial dependency of vegetation-environment linkages in an anthropogenetically influenced wetland ecosystem. *Ecosystems,* **7(1),** 75–97.

Kinzig, A.P., P. Ryan, M. Etienne, H. Allison, T. Elmqvist, and B.H. Walker, 2006. Resilience and regime shifts: Assessing cascading effects. *Ecology and Society,* **11(1),** 20.

Klein, Eric, E.E. Berg, and Roman Dial, 2005. Wetland drying and succession across the Kenai Peninsula lowlands, south-central Alaska. *Canadian Journal of Forest Research,* **35,** 1931–1942.

Kleypas, J.A., J.W. McManus, and L.A.B. Meñez, 1999. Environmental limits to coral reef development: Where do we draw the line? *American Zoologist,* **39,** 146–159.

Kleypas, J.A., R.W. Buddemeier, and J.-P. Gattuso, 2001. The future of coral reefs in an age of global change. *International Journal of Earth Sciences*, **90**, 426–437.

Knowlton, Nancy, 1992. Thresholds and multiple stable states in coral reef community dynamics. *American Zoologist,* **32(6)**, 674–682.

Kurz, W.A., Graham Stinson, Gregory J. Rampley, Caren C. Dymond, and Eric T. Nelson, 2008. Risk of natural disturbances makes future contribution of Canada's forests to the global carbon cycle highly uncertain. *Proceedings of the National Academy of Sciences,* **105(5)**, 1551–1555.

Kurz, W.A., and M.J. Apps, 1999. A 70-year retrospective analysis of carbon fluxes in the Canadian forest sector. *Journal of Applied Ecology*, **9(2)**, 526–547.

Laird, K.R., and B.F. Cumming, 1998. Tracing droughts into the past. *Science Spectra*, **11**, 50–57.

Lake, P.S., 2003: Ecological effects of perturbation by drought in flowing waters. *Freshwater Biology*, **48**, 1161–1172.

Lee, K.N., 1993. *Compass and Gyroscope: Integrating Science and Politics for the Environment.* Island Press, Washington, D.C.

Lemieux, C.J., and D.J. Scott, 2005. Climate change, biodiversity conservation and protected area planning in Canada. *Canadian Geographer,* **49(4)**, 384–397.

Lessios, H.A., D.R. Robertson, and J.D. Cubit, 1984. Spread of *Diadema* mass mortality through the Caribbean. *Science*, **226**, 335–337.

Lewontin, R.C., 1969. The meaning of stability. *Symposia in Biology,* **22**, 13–24.

Light, H.M., M.R. Darst, and J.W. Grubbs, 1998. *Aquatic Habitats in Relation to River Flow in the Apalachicola River Floodplain, Florida.* U.S. Geological Survey Professional Paper 1594.

Lite, S.J., and J.C. Stromberg, 2005. Surface water and ground-water thresholds for maintaining *Populus-Salix* forests, San Pedro River, Arizona. *Biological Conservation*, **125**, 153–167.

Litzow, M.A., J.D. Urban, and B.J. Laurel, 2008. Increased spatial variance accompanies reorganization of two continental shelf ecosystems. *Ecological Applications*, **18**, 1331–1337.

Lloyd, A.H., and C.L. Fastie, 2003. Recent changes in tree line forest distribution and structure in interior Alaska. *Ecoscience*, **10(2)**, 176–185.

Lloyd, A.L., 2009. Sensitivity of model-based epidemiological parameter estimation to model assumptions. In: *Mathematical and Statistical Estimation Approaches in Epidemiology* [G. Chowell, M. Hyman, L. Bettencourt, and C. Castillo-Chavez (eds.)]. Springer, 123–141.

Logan, J.A., Peter White, Barbara Bentz, and J.A. Powell, 1998. Model analysis of spatial patterns in mountain pine beetle outbreaks. *Theoretical Population Biology,* **53(3)**, 236–255.

Loughlin, T.R., I.N. Sukhanova, E.H. Sinclair, and R.C. Ferrero, 1999. Summary of biology and ecosystem dynamics in the Bering Sea. In: *Dynamics of the Bering Sea* [T.R. Loughlin and K. Ohtani (eds.)]. University of Alaska sea grant AK–SG–99–03, 387–407.

Lovett, G.M., K.C. Weathers, and W.V. Sobczak, 2000. Nitrogen saturation and retention in forested watersheds of the Catskill Mountains, New York. *Journal of Applied Ecology*, **10**, 73–84.

Lovett, G.M., L.M. Christenson, P.M. Groffman, C.G. Jones, J.E. Hart, and M.J. Mitchell, 2002. Insect defoliation and nitrogen cycling in forests. *BioScience,* **52**, 335–341.

Lovvorn, J.R., S.E. Richman, J.M. Grebmeier, and L.W. Cooper, 2003. Diet and body condition of spectacled eiders wintering in pack ice of the Bering Sea. *Polar Biology,* **26**, 259–267.

Ludwig, D., D.D. Jones, and C.S. Holling, 1978. Qualitative analysis of insect outbreak systems: The spruce budworm and forest. *Journal of Animal Ecology,* **47(1)**, 315–332.

Ludwig, J.A., B.P. Wilcox, D.D. Breshears, D.J. Tongway, and A.C. Imeson, 2005. Vegetation patches and runoff-erosion as interacting ecohydrological processes in semiarid landscapes. *Ecology,* **86(2)**, 288–297.

Ludwig, J.A., Wiens, J.A., Tongway, D.J, 2000. A scaling rule for landscape patches and how it applies to conserving soil resources in savannas. *Ecosystems,* 3, 84–97.

Macklin, S.A., and G.L. Hunt, Jr. (eds.), 2004. *The Southeast Bering Sea Ecosystem: Implications for Marine Resource Management.* National Oceanographic and Atmospheric Administration Coastal Ocean Program Decision Analysis Series No. 24.

Margalef, R., 1963. On certain unifying principles in ecology. *American Naturalist,* **97(897)**, 357–374.

Matthews, W.J., and Edie Marsh-Matthews, 2003. Effects of drought on fish across axes of space, time and ecological complexity. *Freshwater Biology,* **48(7)**, 1232–1253.

May, R.M., 1973. *Stability and Complexity in Model Ecosystems.* Princeton University Press.

May, R.M., 1977. Thresholds and breakpoints in ecosystems with a multiplicity of stable states. *Nature,* **269(5628)**, 471–477.

May, R.M., S.A. Levin, and George Sugihara, 2008. Complex systems: Ecology for bankers. *Nature,* **451**, 893–895.

May, Robert, and Angela McLean, 2007. *Theoretical Ecology,* third edition. Oxford University Press, New York.

McCann, K.S., 2000. The diversity-stability debate. *Nature,* **405**, 228–233.

McClanahan, T.R., Maina, J., Moothien-Pillay, R., and A.C. Baker, 2005. Effects of geography, taxa, water flow, and temperature variation on coral bleaching intensity in Mauritius. *Marine Ecology Progress Series,* **298**, 131–142.

McDowell, N., W.T. Pockman, C.D. Allen, D.D. Breshears, N. Cobb, T. Kolb, J. Sperry, A. West, D. Williams, E.A. Yepez, 2008. Mechanisms of plant survival and mortality during drought: Why do some plants survive while others succumb to drought? Tansley Review, *New Phytologist,* doi:10.1111/j.1469–8137.2008.02436.x.

McKenzie, D., and C.D. Allen, 2007. Meetings: Climate change and disturbance interactions: Workshop on climate change and disturbance interactions in western North America, Tucson, AZ, 12–15 February 2007. *Eos, Transactions, AGU,* **88(21)**, 227.

Michener, W.K., T.J. Baerwald, Penelope Firth, M.A. Palmer, J.L. Rosenberger, E.A. Sandlin, and Herman Zimmerman, 2001. Defining and unraveling biocomplexity. *BioScience,* **51(12)**, 1018–1023.

Miles, E.L., A.K. Snover, A.F. Hamlet, B. Callahan, and D. Fluharty, 2007. Pacific Northwest Regional Assessment: The impacts of climate variability and climate change on the water resources of the Columbia River Basin. *Journal of American Water Resources Association,* **36**, 399–420.

Millar, C.I., in press. Climate change; Confronting the global experiment. In: *Proceedings of the 27th Annual Forest Vegetation Management Conference, Growing the Future* [S. Cooper and S. Frederickson (eds.)]. January 17–19, 2006, Redding, CA. University of California, Shasta County Cooperative Extension, Redding.

Millar, C.I., N.L. Stephenson, and S.L. Stephens, 2007. Climate change and forests of the future: Managing in the face of uncertainty. *Ecological Applications,* **17**, 2145–2151.

Millenium Ecosystem Assessment, 2005. *Ecosystems and Human Well-Being: Biodiversity Synthesis.* World Resources Institute, Washington, D.C.

Miller, W.D., S.C. Neubauer, and I.C. Andereson, 2001. Effects of sea level induced disturbances on high salt marsh metabolism. *Estuaries,* **24(3)**, 357–367.

Milne, B.T., 1998. Motivation and benefits of complex systems approaches in ecology. *Ecosystems,* **1(5)**, 449–456.

Mitchell, M.J., C.T. Driscoll, J.S. Khal, G.E. Likens, P.S. Murdoch, and L.H. Pardo, 1996. Climatic control of nitrate loss from forested watersheds in the northeast United States. *Environmental Science and Technology,* **30(8)**, 2609–2612.

Moody, J.A., and D.A. Martin, [in press]. Synthesis of sediment yields after wildland fire in different rainfall regimes in the Western United States. *International Journal of Wildland Fire.*

Moorcroft, P.R., G.C. Hurtt, and S.W. Pacala, 2001. A method for scaling vegetation dynamics: The ecosystem demography model (ED). *Ecological Monographs,* **71(4)**, 557–586.

Moore, S.E., J.M. Grebmeier, and J.R. Davis, 2003. Gray whale distribution relative to forage habitat in the northern Bering Sea: Current conditions and retrospective summary. *Canadian Journal of Zoology,* **81**, 734–742.

Mueter, F.J., and M.A. Litzow, 2008. Sea-ice retreat alters the biogeography of the Bering Sea continental shelf. *Ecological Applications,* **18**, 309–320.

Mulholland, P.J., B.J. Roberts, W.R. Hill, and J.G. Smith, [in press]. Stream ecosystem responses to the 2007 spring freeze in the southeastern United States: Unexpected effects of climate change. *Global Change Biology.*

Muradian, Roldan, 2001. Ecological thresholds: A survey. *Ecological Economics,* **38(1),** 7–24.

Murdoch, Thaddeus, Annie Glasspool, and Mike Colella, 2008. The results of the first comprehensive benthic assessment of the coral reef habitats of Bermuda. In: *11th International Coral Reef Symposium,* Fort Lauderdale, FL, July 7–11, 2008, abstract, 451. National Coral Reef Institute, Fort Lauderdale.

NRC, 2002. *Interim Report From the Committee on Endangered and Threatened Fishes in the Klamath River Basin: Scientific Evaluation of Biological Opinions on Endangered and Threatened Fishes in the Klamath River Basin.* National Academy Press, Washington, D.C.

Neilson, R.P., 1993. Transient ecotone response to climatic change: Some conceptual and modeling approaches. *Ecological Applications,* **3(3),** 385–395.

Neitzel, D.A., M.J. Scott, S.A. Shankle and J.C. Chatters, 1991. The effect of climate change on stream environments: The salmonid resource of the Columbia River Basin. *Northwest Environmental Journal,* 7, 271–293.

Nelson, K.C., and M.A. Palmer, 2007. Stream temperature surges under urbanization and climate change: Data, models, and responses. *Journal of the American Water Resources Association,* 43, 440–452.

Newmaster, S.G., W.C. Parker, Wayne Bell, and J.M. Paterson, 2007. Effects of forest floor disturbances by mechanical site preparation on floristic diversity in a central Ontario clearcut. *Forest Ecology and Management,* **246(2–3),** 196–207.

Nkemdirim, Lawrence, and Lena Weber, 1999. Comparison between the drought of the 1930's and the 1980's in the southern prairies of Canada. *Journal of Climate,* **12(8),** 2434–3450.

Obst, B.S., R.W. Russell, G.L. Hunt, Z.A. Eppley, and N.M. Harrison, 1995. Foraging radii and energetics of least auklets (*Aethia pusilla*) breeding on three Bering Sea islands. *Physiological Zoology,* **68,** 647–672.

Odum, E.P., 1969. The strategy of ecosystem development. *Science,* **164(3877),** 262–270.

Ojima, D.S., and J.M. Lackett, 2002. *Preparing for a Changing Climate: The Potential Consequences of Climate Variability and Change—Central Great Plains.* Report for the U.S. Global Change Research Program. Colorado State University.

Omernik, J.M., 1987. Ecoregions of the conterminous United States. Map (scale 1:7,500,000). *Annals of the Association of American Geographers* 77(1):118–125.

Omernik, J.M., 1995. Ecoregions: A spatial framework for environmental management. In: *Biological Assessment and Criteria: Tools for Water Resource Planning and Decision Making* [W.S. Davis and T.P. Simon (eds.)]. Lewis Publishers, Boca Raton, FL, 49–62.

O'Neill, R.V., D.L. DeAngelis, J.B. Waide, and T.F.H. Allen, 1986. *A Hierarchical Concept of Ecosystems.* Princeton University Press.

Osterkamp, T.E., 2007. Characteristics of the recent warming of permafrost in Alaska. *Journal of Geophysical Research,* **112,** F02S02, doi:10.1029/2006JF000578.

Osterkamp, T.E., and V.E. Romanovsky, 1999. Evidence for warming and thawing of discontinuous permafrost in Alaska. *Permafrost and Periglacial Processes,* **10,** 17–37.

Osterkamp, T.E., D.C. Esch, and V.E. Romanovsky, 1997. Infrastructure: Effects of climatic warming on planning, construction and maintenance. In: *Proceedings of the BESIS Workshop.* University of Alaska, Fairbanks, AK, 115–127.

Osterkamp, T.E., L. Viereck, Y. Shur, M.T. Jorgenson, C.H. Racine, A.P. Doyle, and R.D. Boone, 2000. Observations of thermokarst and its impact on boreal forests in Alaska. *Arctic, Antarctic, and Alpine Research,* **32(3),** 303–315.

Overland, J.E., and P.J. Stabeno, 2004. Is the climate of the Bering Sea warming and affecting the ecosystem? *Eos, Transactions, AGU,* **85,** 309–316.

Paulay, G., 1997. Diversity and distribution of reef organisms. In: *Life and Death of Reefs* [Charles Birkeland (ed.)]. Chapman and Hall, 298–353.

Palmer, M.A., C.A. Reidy, Christer Nilsson, Martina Flörke, Joseph Alcamo, P.S. Lake, and Nick Bond, 2007. Climate change and the world's river basins: Anticipating management options. *Frontiers in Ecology and the Environment,* **6(2),** 81–89.

Palmer, M.A., J.D. Allan, and C.A. Butman, 1996. Dispersal as a regional process affecting the local dynamics of marine and stream benthic invertebrates. *Trends in Ecology and Evolution,* **11(8),** 322–326.

Parker, W.C., S.J. Colombo, M.L. Cherry, M.D. Flannigan, Sylvia Greifenhagen, Chris Papadopol, M.D. Flannigan, R.S. McAlpine, and Taylor Scarr, 2000. Third millennium forestry: What climate change might mean to forests and forest management in Ontario. *Forestry Chronicle,* **76(3)**, 445–463.

Parmesan, Camille, and Hector Galbraith, 2004. *Observed Impacts of Global Climate Change in the United States.* Pew Center on Global Climate Change, Washington, D.C.

Parr, T.W., A.R.J. Sier, R.W. Battarbee, A. Mackay, and J. Burgess, 2003. Detecting environmental change: Science and society—Perspectives and long-term research and monitoring in the 21st century. *Science of the Total Environment,* **310(1–3)**, 1–8.

Pascual, Mercedes, and S.P. Ellner, 2000. Linking ecological patterns to environmental forcing via nonlinear time series models. *Ecology,* **81(10)**, 2767–2780.

Paul, M.J., and J.L. Meyer, 2001. Streams in the urban landscape. *Annual Review* of *Ecology, Evolution, and Systematics,* **32**, 333–365.

Payne, J.T., A.W. Wood, A.F. Hamlet, R.N. Palmer and D.P. Lettenmaier, 2004. Mitigating the effects of climate change on the water resources of the Columbia River Basin. *Climatic Change,* **62**, 233–256.

Peters, D.P.C., B.T. Bestelmeyer, and M.G. Turner, 2007. Cross-scale interactions and changing pattern-process relationships: Consequences for system dynamics. *Ecosystems,* **10**, 790–796.

Peters, D.P.C., B.T. Bestelmeyer, J.E. Herrick, E.L. Fredrickson, H.C. Monger, and K.M. Havstad, 2006. Disentangling complex landscapes: New insights into arid and semiarid system dynamics. *BioScience,* **56**, 491–501.

Peters, D.P.C., R.A. Pielke, Sr., B.T. Bestelmeyer, C.D. Allen, Stuart Munson-McGee, and K.M. Havstad, 2004. Cross-scale interactions, nonlinearities, and forecasting catastrophic events. *Proceedings of the National Academy of Sciences,* **101(42)**, 15130–15135.

Peters, D.P.C., P.M Groffman, K.J. Nadelhoffer, N.B. Grimm, S.L. Collins, W.K. Michener, and M.A. Huston, 2008. Living in an increasingly connected world: A framework for continental-scale environmental science. *Frontiers in Ecology and the Environment,* **6(5)**, 229–237.

Piatt, J.F., and A.M. Springer, 2003. Advection, pelagic food webs and the biogeography of seabirds in Beringia. *Marine Ornithology,* **31**, 141–154.

Piatt, J.F., S.A. Hatch, B.D. Roberts, W.W. Lidster, J.L. Wells, and J.C. Haney, 1988. *Populations, Productivity, and Feeding Habits of Seabirds on St. Lawrence, Alaska.* U.S. Fish and Wildlife Service OCS Study MMS 88–0022.

Pielke, R.A., Sr., T.J. Lee, J.H. Copeland, J.L. Eastman, C.L. Ziegler, and C.A. Finley, 1997. Use of USGS-provided data to improve weather and climate simulations. *Journal of Applied Ecology,* **7(1)**, 3–21.

Pimm, S.L., 1984. The complexity and stability of ecosystems. *Nature,* **307**, 321–326.

Poff, N.L., J.D. Allan, M.B. Bain, J.R. Karr, K.L. Prestegaard, B.D. Richetr, R.E. Sparks, and J.C. Stromberg, 1997. The natural flow regime. *BioScience,* **47(11)**, 769–784.

Poiani, K.A., and W.C. Johnson, 1991. Global warming and prairie wetlands. *BioScience,* **41(9)**, 611–618.

Poiani, K.A., and W.C. Johnson, 1993a. A spatial simulation model of hydrology and vegetation dynamics in semi-permanent prairie wetlands. *Ecological Applications,* **3(2)**, 279–293.

Poiani, K.A., and W.C. Johnson, 1993b. Potential effects of climate change on a semi-permanent prairie wetland. *Climatic Change,* **24(3)**, 213–232.

Poiani, K.A., W.C. Johnson, and T.G.F. Kittel, 1995. Sensitivity of a prairie wetland to increased temperature and seasonal precipitation changes. *Journal of the American Water Resources Association,* **31(2)**, 283–294.

Poiani, K.A., W.C. Johnson, G.A. Swanson, and T.C. Winter, 1996. Climate change and northern prairie wetlands: Simulations of long-term dynamics. *Limnology and Oceanography,* **41(5)**, 871–881.

Prato, Tony, 2004. Alleviating multiple threats to protected areas with adaptive ecosystem management. *The George Wright Forum,* **20(4)**, 41–52.

Prato, Tony and Dan Fagre, 2007. Achieving ecosystem sustainability. In: *Sustaining Rocky Mountain Landscapes: Science, Policy and Management of the Crown of the Continent Ecosystem* [Tony Prato and Dan Fagre (eds.)]. Resources for the Future Press, Washington, D.C., 302–311.

Pulwarty, R.S., K. Jacobs, and R. Dole, 2005. The hardest working river: Drought and critical water problems on the Colorado. In: *Drought and Water Crises: Science, Technology, and Management Issues* [D.A. Wilhite (ed.)]. Taylor and Francis Press, 249–285.

Purves, Drew, and Stephen Pacala, 2008. Predictive models of forest dynamics. *Science, 320,* 1452–1453.

Pyke, D.A., J.E. Herrick, Patrick Shaver, and Mike Pellant, 2002. Rangeland health attributes and indicators for qualitative assessment. *Journal of Range Management,* **55(6),** 584–597.

Radford, J.Q., A.F. Bennett, and G.J. Cheers, 2005. Landscape-level thresholds of habitat cover for woodland-dependent birds. *Biological Conservation,* **124(3),** 317–337.

Redman, C.L., and A.P. Kinzig, 2003. Resilience of past landscapes: Resilience theory, society, and the longue durée. *Conservation Ecology,* **7(1),** 14.

Reynolds, J.F., and D.M. Stafford Smith, eds., 2002. *Global Desertification: Do Humans Cause Deserts?.* Dahlem University Press, Berlin, Workshop Report 88.

Rich, P.M., D.D. Breshears, and A.B. White, 2008. Phenology of mixed woody-herbaceous ecosystems following extreme events: Net and differential responses. *Ecology,* **89(2),** 342–352.

Richmond, Robert, *et al.,* 2002. Status of the coral reefs in Micronesia and American Samoa: US affiliated and freely associated islands in the Pacific. In: *Status of Coral Reefs of the World* [Clive Wilkinson (ed.)]. Australian Institute of Marine Science, Townsville, Queensland, Australia, 217–236.

Rietkerk, M., and J. van de Koppel, 1997. Alternate stable states and threshold effects in semi-arid grazing systems. *Oikos,* **79(1),** 69–76.

Rietkerk, Max, S.C. Dekker, P.C. de Ruiter, and Johan van de Koppel, 2004. Self-organized patchiness and catastrophic shifts in ecosystems. *Science,* **305(5692),** 1926–1929.

Riordan, Brian, David Verbyla, and A.D. McGuire, 2006. Shrinking ponds in subarctic Alaska based on 1950–2002 remotely sensed images. *Journal of Geophysical Research Biogeosciences,* **111,** G04002, doi:10.1029/2005JG000150.

Roach, A.T., K. Aagaard, C.H. Pease, S.A. Salo, T. Weingartner, V. Pavlov, and M. Kulakov, 1995. Direct measurements of transport and water properties through the Bering Strait. *Journal of Geophysical Research,* **100(C9),** 18443–18457.

Rodó, Xavier, and F.A. Comín, 2002. *Global Climate: Current Research and Uncertainties in the Climate System.* Springer, New York.

Rosenzweig, Cynthia, and Daniel Hillel, 1993. The dust bowl of the 1930's: Analog of greenhouse effect in the Great Plains? *Journal of Environmental Quality,* **22,** 9–22.

Rosenzweig, M.L., 1971. Paradox of enrichment: Destabilization of exploitation ecosystems in ecological time. *Science,* **171(3969),** 385–387.

Rowan, Rob, 2004. Coral bleaching: Thermal adaptation in reef coral symbionts. *Nature,* **430,** 742.

Rowan, Rob, and Nancy Knowlton, 1995. Intraspecific diversity and ecological zonation in coral-algal symbiosis. *Proceedings of the National Academy of Science,* **92,** 2850–2853.

Roy, A.H., A.D. Rosemond, M.J. Paul, D.S. Leigh, and J.B. Wallace, 2003. Stream macroinvertebrate response to catchment urbanization (Georgia, U.S.A.). *Freshwater Biology,* **48(2),** 329–346.

Russell, R.W., N.M. Harrison, and G.L. Hunt, Jr., 1999. Foraging at a front: Hydrography, zooplankton, and avian planktivory in the northern Bering Sea. *Marine Ecology Progress Series,* **182,** 77–93.

Sabine, S.L., R.A. Feely, Nicolas Gruber, R.M. Key, Kitack Lee, J.L. Bullister, Rick Wanninkhof, C.S. Wong, D.W.R. Wallace, Bronte Tilbrook, F.J. Millero, T.-H. Peng, Alexander Kozyr, Tsueno Ono, and A.F. Rios, 2004. The oceanic sink for anthropogenic CO_2. *Science,* **305(5682),** 367–371.

Sala, Enric, and George Sugihara, 2005. Food-web theory provides guidelines for marine ecosystems. In: *Aquatic Food Webs: An Ecosystem Approach* [Andrea Belgrano, U.M. Scharler, Jennifer Dunne, and R.E. Ulanowicz (eds.)]. Oxford University Press, 170–183.

Sala, Enric, and M.H. Graham, 2002. Community-wide distribution of predator-prey interaction strength in kelp forests. *Proceedings of the National Academy of Science,* **99(6),** 3678–3683.

Savage M., and J.N. Mast, 2005. How resilient are southwestern ponderosa pine forests after crown fires? *Canadian Journal of Forest Research,* **35**, 967–977.

Scheffer, Marten, and S.R. Carpenter, 2003. Catastrophic regime shifts in ecosystems: Linking theory to observation. *Trends in Ecological Evolution,* **18(12)**, 648–656.

Scheffer, Marten, Steve Carpenter, J.A. Foley, Carl Folke, and Brian Walker, 2001. Catastrophic shifts in ecosystems. *Nature,* **413**, 591–596.

Schlesinger, W.H., J.F. Reynolds, G.L. Cunningham, L.F. Huenneke, W.M. Jarrell, R.A. Virginia, and W.G. Whitford, 1990. Biological feedbacks in global desertification. *Science,* **247(4946)**, 1043–1048.

Schmitz, O.J., Eric Post, C.E. Burns, and K.M. Johnston, 2003. Ecosystem responses to global climate change: Moving beyond color mapping. *BioScience,* **53(12)**, 1199–1205.

Schneider, S.H., 2004. Abrupt non-linear climate change, irreversibility and surprise. *Global Environmental Change Part A,* **14(3)**, 245–258.

Scholze, Marco, Wolfgang Knorr, N.W. Arnell, and I.C. Prentice, 2006. A climate-change risk analysis for world ecosystems. *Proceedings of the National Academy of Sciences,* **103**, 13116–13120.

Scott, T.A., 2000. The selection and design of multiple-species habitat preserves. *Environmental Management,* **26**, 37–53.

Scott, D.J., and R. Suffling, 2000. *Climate Change and Canada's National Parks.* Environment Canada, Toronto.

Scott, Daniel, and Christopher Lemieux, 2005. Climate change and protected area policy and planning in Canada. *The Forestry Chronicle,* **81(5)**, 696–703.

Shapley, M.D., W.C. Johnson, D.R. Engstrom, and W.R. Osterkamp, 2005. Late-holocene flooding and drought in the northern Great Plains, USA, reconstructed from tree rings, lake sediments, and ancient shorelines. *The Holocene,* **5(1)**, 29–41.

Skaggs, Richard H., 1975. Drought in the United States, 1931–40. *Annals of the Association of American Geographer,* **65(3)**, 391–402.

Smith, L.C., Y. Sheng, G.M. MacDonald, and L.D. Hinzman, 2005. Disappearing arctic lakes. *Science,* **308(5727)**, 1429.

Smith, L.W., and Charles Birkeland, 2007. Effects of intermittent flow and irradiance level on back reef *Porites* corals at elevated seawater temperatures. *Journal of Experimental Marine Biology and Ecology,* **341(2)**, 282–294.

Smithwick, E.A.H, M.G. Turner, M.C. Mack, and F.S. Chapin III, 2005. Postfire soil N cycling in northern conifer forests affected by severe, stand-replacing wildfires. *Ecosystems,* **8**, 163–181.

Soule, P.T., and P.A. Knapp, 1999. Western juniper expansion on adjacent disturbed and near-relict sites. *Journal of Range Management,* **52**, 525–533.

Springer, A.M., C.P. McRoy, and K.R. Turco, 1989. The paradox of pelagic food webs in the northern Bering Sea—II. Zooplankton communities. *Continental Shelf Research,* **9**, 359–386.

Stabeno, P.J., and J.E. Overland, 2001. Bering Sea shift towards an earlier spring transition. *Eos, Transactions, AGU,* **82**, 317–321.

Stabeno, P.J., G.L. Hunt, Jr., J.M. Napp, and J.D. Schumacher, 2006. Physical forcing of ecosystem dynamics on the Bering Sea shelf. In: *The Sea* [A.R. Robinson and K.H. Brink (eds.)]. The Global Coastal Ocean: Interdisciplinary Regional Studies and Syntheses, Part B. Harvard University Press, John Wiley and Sons, New York, **14**, 1177–1212.

Stabeno, P.J., N.A. Bond, and S.A. Salo, 2007. On the recent warming of the southeastern Bering Sea shelf. *Deep Sea Research II,* **54(23–26)**, 2599–2618.

Stabeno, P.J., N.A., Bond, N.B. Kachel, S.A. Salo, and J.D. Schumacher, 2001. On the temporal variability of the physical environment over the south-eastern Bering Sea. *Fisheries Oceanography,* **10(1)**, 81–98.

Steele, J.H., 1996. Regime shifts in fisheries management. *Fisheries Research,* **25(1)**, 19–23.

Steele, J.H., and E.W. Henderson, 1984. Modeling long-term fluctuations in fish stocks. *Science,* **224(4652)**, 985–987.

Stephenson, N.L., and P.J. van Mantgem, 2005. Forest turnover rates follow global and regional patterns of productivity. *Ecology Letters,* **8**, 524–531.

Stewart, I.T., D.R. Cayan, and M.D. Dettinger, 2004. Changes in snowmelt runoff timing in western North America under a 'business as usual' climate change scenario. *Climatic Change,* **62(1–3)**, 217–232.

Stoddard, J.L., 1994. Long-term changes in watershed retention of N. In: *Chemistry of Lakes and Reservoirs* [L.A. Baker (ed.)]. Advances in Chemistry Society Series No. 237, Washington, D.C., 223–284.

Stoddard, J.L., D.S. Jeffries, A. Lükewille, T.A. Clair, P.J. Dillon, C.T. Driscoll, M. Frosius, M. Johannessen, J.S. Kahl, J.H. Kellogg, A. Kemp, J. Mannio, D.T. Monteith, P.S. Murdoch, S. Patrick, A. Rebsdorf, B.L. Skjelvåle, M.P. Stainton, T. Traaen, H. van Dam, K.E. Webster, J. Wieting, and A. Wilander, 1999. Regional trends in aquatic recovery from acidification in North America and Europe. *Nature,* **401,** 575–578.

Stoffels, R.J., K.R. Clarke, and G.P. Closs, 2005. Spatial scale and benthic community organization in the littoral zones of large oligotrophic lakes: Potential for cross-scale interactions. *Freshwater Biology,* **50(7),** 1131–1145.

Stone, R.S., E.G. Dutton, J.M. Harris, and D. Longnecker, 2002. Earlier spring snowmelt in northern Alaska as an indicator of climate change. *Geophysical Research,* **107,** 4089.

Stringham, T.K., W.C. Krueger, and P.L. Shaver, 2003. State and transition modeling: An ecological process approach. *Journal of Range Management,* **56,** 106–113.

Stroeve, J., M.C. Serreze, F. Fetterer, T. Arbetter, W. Meier, J. Maslanik, and K. Knowles, 2005. Tracking the Arctic's shrinking ice cover; another extreme September sea-ice minimum in 2004. *Geophysical Research Letters,* **32(L04501),** 3, 5, doi:10.1029/2004GL021810.

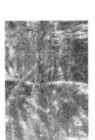

Stromberg, J.C., K.J. Bagstad, J.M. Leenhouts, S.J. Lite, and Elizabeth Makings, 2005. Effects of stream flow intermittency on riparian vegetation of a semiarid region river (San Pedro River, Arizona). *River Research and Applications,* **21(8),** 925–938.

Stromberg, J.C., V.B. Beauchamp, M.D. Dixon, S.J. Lite, and C. Paradzick, 2007. Importance of low-flow and high-flow characteristics to restoration of riparian vegetation along rivers in arid south-western United States. *Freshwater Biology,* **52(4),** 651–679.

Sturm, Matthew, Charles Racine, and Kenneth Tape, 2001. Increasing shrub abundance in the Arctic. *Nature,* **411,** 546–547.

Suarez, M.L., Luciana Ghermandi, and Thomas Kitzberger, 2004. Factors predisposing episodic drought-induced tree mortality in Nothofagus—site, climatic sensitivity and growth trends. *Journal of Ecology,* **92(6),** 954–966.

Suding, K.N., K.L. Gross, and G.R. Houseman, 2004. Alternative states and positive feedbacks in restoration ecology. *Trends in Ecology and Evolution,* **19(1),** 46–53.

Sugihara, George, 1984. Graph theory, homology, and food webs. *Proceedings of Symposia in Applied Mathematics,* **30,** 83–89.

Sugihara, George, 1994. Nonlinear forecasting for the classification of natural time series. *Philosophical Transactions: Physical Sciences and Engineer*ing, **348(1688),** 477–495.

Sugihara, George, and R.M. May, 1990. Nonlinear forecasting as a way of distinguishing chaos from measurement error in time series. *Nature,* **344,** 737–741.

Sutherland, J.P., 1974. Multiple stable points in natural communities. *American Naturalist,* **108,** 859.

Swank, W.T., and J.M. Vose, 1997. Long-term nitrogen dynamics of Coweeta forested watersheds in the southeastern United States of America. *Global Biogeochemical Cycles,* **11(4),** 657–671.

Swetnam, T.W., and J.L. Betancourt, 1998. Mesoscale disturbance and ecological response to decadal climatic variability in the American southwest. *Journal of Climate,* **11,** 3128–3147.

Tallmon, D.A., E.S. Jules, N.J. Radke, and L.S. Mills, 2003. Of mice and men and trillium: Cascading effects of forest fragmentation. *Ecological Applications,* **13(5),** 1193–1203.

Tape, Ken, Matthew Sturm, and Charles Racine, 2006. The evidence for shrub expansion in Northern Alaska and the Pan-Arctic. *Global Change Biology,* **12,** 686–702.

Thom, Rene, 1975. *Structural Theory and Morphogenesis: An Outline of a General Theory of Models.* W.A. Benjamin, Reading, MA.

Thom, Rene, 1971. Modern mathematics': An educational and philosophical error? *American Scientist,* **59(6),** 695–699.

Thuiller, Wilfried, Sandra Lavorel, M.B. Araujo, M.T. Sykes, and I.C. Prentice, 2005. Climate change threats to plant diversity in Europe. *Proceedings of the National Academy of Sciences,* **102,** 8245–8250.

Tilman, David, and David Wedin, 1991. Oscillations and chaos in the dynamics of a perennial grass. *Nature,* **353,** 653–655.

Turner, B.L., II, R.E. Kasperson, P.A. Matson, J.J. McCarthy, R.W. Corell, Lindsey Christenses, Noelle Eckley, J.X. Kasperson, Amy Luers, M.L. Martello, Colin Polsky, Alexander Pulsipher, and Andrew Schiller, 2003. A framework for vulnerability analysis in sustainable science. *Proceedings of the National Academy of Sciences,* **100(14),** 8074–8079.

Urban, D.L., 2005. Modeling ecological processes across scales. *Ecology,* **86(8),** 1996–2006.

U.S. Forest Service, 2006. *Forest Insect and Disease Conditions in the Southwestern Region, 2005.* U.S. Forest Service publication PR–R3–16–1.

van der Valk, A.G. (ed.), 1989. *Northern Prairie Wetlands.* Iowa State University Press, Ames, Iowa.

van Mantgem, P.J., and N.L. Stephenson, 2007. Apparent climatically-induced increase of tree mortality rates in a temperate forest. *Ecology Letters,* **10,** 909–916.

van Mantgem, P.J., N.L. Stephenson, J.C. Byrne, L.D. Daniels, J.F. Franklin, P.Z. Fulé, M.E. Harmon, A.J. Lasrson, J.M. Smith, A.H. Taylor, and T.T. Veblen, [in review]. Widespread increase of tree mortality rates in the western United States. *Science.*

Van Nes, E.H., and M. Scheffer, 2007. Slow recovery from perturbations as a generic indicator of a nearby catastrophic shift. *The American Naturalist,* **169(6),** 738.

Veenhuis, J., 2002. *Effects of Wildfire on the Hydrology of Capulin and Rito de los Frijoles Canyons, Bandelier National Monument, New Mexico.* U.S. Geological Survey Water-Resources Investigations Report 02–4152.

Vitousek, P.M., 1994. Beyond global warming: Ecology and global change. *Ecology,* **75(7),** 1861–1876.

Voldseth, R.A., W.C. Johnson, T. Gilmanov, G.R. Guntenspergen, and B. Millett, 2007. Model estimation of land-use effects on water levels of northern prairie wetlands. *Ecological Applications,* **17(2),** 527–540.

Vorosmarty, C.J., Pamela Green, Joseph Salisbury, and R.B. Lammers, 2000. Global water resources: Vulnerability from climate change and population growth. *Science,* **289(5477),** 284–288.

Walker, Brian, and J.A. Meyers, 2004. Thresholds in ecological and social-ecological systems: A developing database. *Ecology and Society,* **9(2),** 3.

Walker, Brian, C.S. Holling, S.R. Carpenter, and Ann Kinzig, 2004. Resilience, adaptability and transformability in social-ecological systems. *Ecology and Society,* **9(2),** 5.

Walker, Brian, Stephen Carpenter, John Anderies, Nick Abel, G.S. Cumming, Marco Janssen, Louis Lebel, Jon Norberg, G.D. Peterson, and Rusty Pritchard, 2002. Resilience management in social-ecological systems: A working hypothesis for a participatory approach. *Conservation Ecology,* **6(1),** 14.

Walker, M.D., C.H. Wahren, R.D. Hollister, G.H.R. Henry, L.E. Ahlquist, J.M. Alatalo, M.S. Bret-Harte, M.P. Calef, T.V. Callaghan, A.B. Carroll, H.E. Epstein, I.S. Jónsdóttir, J.A. Klein, Borgþór Magnússon, Ulf Molau, S.F. Oberbauer, S.P. Rewa, C.H. Robinson, G.R. Shaver, K.N. Suding, C.C. Thompson, Anne Tolvanen, Ørjan Totland, P.L. Turner, C.E. Tweedie, P.J. Webber, and P.A. Wookey, 2006. Plant community responses to experimental warming across the tundra biome. *Proceedings of the National Acamemy of Sciences,* **103(5),** 1342–1346.

Walters, Carl, 1986. *Adaptive Management of Renewable Resources.* Macmillan and Co., New York.

Wamelink, G.W.W., ter Braak, C.J.F., and van Dobben, H.F, 2003. Changes in large-scale patterns of plant biodiversity predicted from environmental economic scenarios. *Landscape Ecology,* **18,** 513–527.

Watson, R.T., et al., 1996. Summary for policymakers. In: *Climate Change 1995: Impacts, Adaptations, and Mitigation of Climate Change: Scientific-Technical Analyses: Contribution of Working Group II to the Second Assessment Report of the Intergovernmental Panel on Climate Change* [R.T. Watson, M.C. Zinyowera, and R.H. Moss (eds.)]. Cambridge University Press, Cambridge, 1–18.

Welch, David, 2005. What should protected areas managers do in the face of climate change? *George Wright Forum,* **22(1),** 75–93.

Weller, M.W., 1965. Chronology of pair formation in some nearctic *Aythya* (Anatidae). *The Auk,* **82(2),** 227–235.

Westerling, A.L., H.G. Hidalgo, D.R. Cayan, and T.W. Swetnam, 2006. Warming and earlier spring increase western U.S. forest wildfire activity. *Science,* **313,** 940–943.

Wiens, J.A., 1989. Spatial scaling in ecology. *Functional Ecology,* **3(4)**, 385–397.

Wilcox, B.P., D.D. Breshears, and C.D. Allen, 2003. Eco-hydrology of a resource-conserving semiarid woodland: Temporal and spatial scaling and disturbance. *Ecological Monographs,* **73(2)**, 223–239.

Wilhere, G.F., 2002. Adaptive management in habitat conservation plans. *Conservation Biology,* **16(1)**, 20–29.

Winter, T.C., 2000. The vulnerability of wetlands to climate change: A hydrologic landscape perspective. *Journal of the American Water Resources Assoiation,* **36(2)**, 305–311.

Winter, T.C., and D.O. Rosenberry, 1998. Hydrology of prairie pothole wetlands during drought and deluge: A 17-year study of the cottonwood lake wetland complex in North Dakota in the perspective of longer term measured and proxy hydrological records. *Climatic Change,* **40(2)**, 189–209.

Winter, T.C., and M.K. Woo, 1990. Hydrology of lakes and wetlands. In: *Surface Water Hydrology* [M.G. Wolman and H.D. Riggs (eds.)]. Geological Society of America, Boulder, CO, **O–1**, 159–187.

Wissel, C., 1984. A universal law of the characteristic return time near thresholds. *Oecologia,* **65(1)**, 101–107.

Woodhouse, C.A., and J.T. Overpeck, 1998. 200 years of drought variability in the central United States. *Bulletin of the American Meteorological Society,* **79**, 2693–2714.

Worm, Boris, and R.A. Meyers, 2003. Meta-analysis of cod-shrimp interactions reveals top-down control in oceanic food webs. *Ecology,* **84**, 162–173.

Xenopoulos, M.A., D.M. Lodge, Joseph Alcamo, Michael Märker, Kerstin Schulze, and D.P. Van Vuuren, 2005. Scenarios of freshwater fish extinctions from climate change and water withdrawal. *Global Change Biology,* **11**, 1557–1564.

Yao, Jin, D.P.C. Peters, K.M. Havstad, R.P. Gibbens, and J.E. Herrick, 2006. Multi-scale factors and long-term responses of Chihuahuan desert grasses to drought. *Landscape Ecology,* **21(8)**, 1217–1231.

Yoshikawa, Kenji, and L.D. Hinzman, 2003. Shrinking thermokarst ponds and groundwater dynamics in discontinuous permafrost near Council, Alaska. *Permafrost and Perglacial Processes,* **14(2)**, 151–160.

Zeller, Dirk, Shawn Booth, and Daniel Pauly, 2006b. Fisheries contributions to the gross domestic product: Underestimating small-scale fisheries in the Pacific. *Marine Resource Economics,* **21(4)**, 355–374.

Zeller, Dirk, Shawn Booth, Peter Craig, and Daniel Pauly, 2006a. Reconstruction of coral reef fisheries catches in American Samoa, 1950–2002. *Coral Reefs,* **25**, 144–152.

Zou, C.B., G.A. Barron-Gafford, and D.D. Breshears, 2007. Effects of topography and woody plant canopy cover on near-ground solar radiation: Relevant energy inputs for ecohydrology and hydropedology. *Geophysical Research Letters,* **34**, L24S21, doi:10.1029/2007GL031484.

Contact Information

Global Change Research Information Office
c/o Climate Change Science Program Office
1717 Pennsylvania Avenue, NW
Suite 250
Washington, DC 20006
202-223-6262 (voice)
202-223-3065 (fax)

The Climate Change Science Program
incorporates the U.S. Global Change Research
Program and the Climate Change Research
Initiative.

To obtain a copy of this document, place
an order at the Global Change Research
Information Office (GCRIO) web site:
http://www.gcrio.org/orders

Climate Change Science Program and the Subcommittee on Global Change Research

William Brennan, Chair
Department of Commerce
National Oceanic and Atmospheric Administration
Director, Climate Change Science Program

Jack Kaye, Vice Chair
National Aeronautics and Space Administration

Allen Dearry
Department of Health and Human Services

Anna Palmisano
Department of Energy

Mary Glackin
National Oceanic and Atmospheric Administration

Patricia Gruber
Department of Defense

William Hohenstein
Department of Agriculture

Linda Lawson
Department of Transportation

Mark Myers
U.S. Geological Survey

Timothy Killeen
National Science Foundation

Patrick Neale
Smithsonian Institution

Jacqueline Schafer
U.S. Agency for International Development

Joel Scheraga
Environmental Protection Agency

Harlan Watson
Department of State

EXECUTIVE OFFICE AND OTHER LIAISONS

Robert Marlay
Climate Change Technology Program

Katharine Gebbie
National Institute of Standards & Technology

Stuart Levenbach
Office of Management and Budget

Margaret McCalla
Office of the Federal Coordinator for Meteorology

Rob Rainey
Council on Environmental Quality

Daniel Walker
Office of Science and Technology Policy